One Soldier's Story

BOB DOLE

One Soldier's Story

· · ·

A MEMOIR

HarperLargePrint
An Imprint of HarperCollins*Publishers*

HarperCollins books may be purchased for educational, business, or sales promotional use. For information, please write: Special Markets Department, HarperCollins Publishers Inc., 10 East 53rd Street, New York, NY 10022.

Grateful acknowledgment is made for permission to reprint from "You'll Never Walk Alone," by Richard Rodgers and Oscar Hammerstein II, copyright © 1945 by Williamson Music, copyright renewed; international copyright secured; all rights reserved; used by permission.

Unless otherwise noted, all photographs are courtesy of the Robert J. Dole Institute of Politics, University of Kansas.

FIRST LARGE PRINT EDITION

Designed by Nancy B. Field

Map by Paul Pugliese

Printed on acid-free paper

Library of Congress Cataloging-in-Publication Data is available upon request.

ISBN 0-06-078714-7

05 06 07 08 09 BVG/RRD 10 9 8 7 6 5 4 3 2 1

This Large Print Book carries the☐
Seal of Approval of N.A.V.H.

Dedication

Most books are dedicated to people the reader doesn't know and may never meet, so I dedicate this book to somebody you know well—a mother or father, grandmother or grandfather who lived through the World War II years—a person of character who has not simply pointed the way through life, but has helped you to find your own direction.

I dedicate this book to the memory of my parents, who showed me the way, and to the men in our family who served their country honorably during World War II—my brother Kenny, who served in the Army Medical Corps; and my brothers-in-law Larry Nelson (Air Force) and Tom Steele (Navy), who are no longer with us; and my brother-in-law John Hanford (Navy), who at eighty-one years is still making a difference.

I also dedicate this book to all the young men and women who have served or are now serving America in our armed forces; and to their families, who know the meaning of sacrifice.

Finally, I dedicate this book to the memory of Sergeant Craig L. Nelson, and to his family, his many friends, and to his unit (Headquarters Company of the 1/156 Armor Battalion of the 256th Infantry Battalion, Louisiana Army National Guard), as representative of all the young men and women who have given their lives in America's fight against terrorism, so we might live in freedom. And to Major Tammy Duckworth (Headquarters Company, 1st Battalion, 106 Aviation, Illinois National Guard), a brave woman wounded in Iraq, who represents all those with their own battles ahead of them, and their own stories to tell.

Acknowledgments

Whether one is competing in sports, ascending a treacherous mountaintop in Italy during World War II, or running for public office, it takes a team effort to succeed. Having the right people in key positions makes all the difference.

Producing this book, **One Soldier's Story**, has certainly reminded me of that fact. It may be only one soldier's story, but it took many people to bring it to fruition. I am deeply grateful to each and every one.

Thanks to Jane Friedman, president and CEO of HarperCollins Publishers, for recognizing that my story represents an entire generation who endured World War II, and for seeing the need to pass it on. Thanks, too, to Tim Duggan, executive editor at HarperCollins, who patiently oversaw this project.

Special thanks to Ken Abraham, who pulled all the pieces of the puzzle together and helped me get my thoughts and feelings out of my memory banks and into print.

Thanks to Mike Marshall, my communications director, who made certain that our story is historically accurate, and that the contributions from every member of the team were carefully evaluated.

Richard Norton Smith, a brilliant writer and thinker, provided enormous editorial assistance as well as his unique perspectives on my story.

Thanks to Doug MacKinnon for proposing his ideas for this book and for his counsel along the way.

Kerry Tymchuk read and reread the manuscript, offering key insights and many good stories that have helped shape these pages.

Jean Bischoff, senior archivist at the Robert J. Dole Institute of Politics on the campus of the University of Kansas, provided invaluable assistance during the research process. Of special note, Jennifer Wade spent countless hours transcribing, categorizing, and archiving several hundred handwritten, World War II letters written by my family members, friends, and me.

Judy Sweets and Linda Kay Pritchard also were most helpful.

My dear sisters, Norma Jean Steele and Gloria Nelson, uncovered these priceless letters and brought them to our attention. Their help in preparing this story for publication has been appreciated by every person on our team.

Thanks to Dr. Charles "Chuck" Peck, my friend and personal medical adviser, who checked the medical portions of the book for accuracy.

Thanks also to Albert Nencioni, one of the few men still alive who was fighting nearby the morning I was wounded, for reading the rough draft of the manuscript and offering valuable suggestions to ensure that our details about the 10th Mountain Division are correct.

Thanks also to Lana Daleo, my personal assistant, for her help in facilitating our many logistical details. And to Polly Walker, Ruth Ann Komarek, Mo Taggart, and Joyce McCluney, at Alston & Bird, for being there whenever we needed their help.

To my daughter, Robin, who jogged my memory, and reminded me of how I coped with my disability, and to her mother, Phyllis, who was a pillar of strength over the years.

And of course, thanks to my friend and attorney Bob Barnett, and his associates at Williams & Connolly LLP, who represented me and made sure that every contract detail was covered. Literary agent Mark Sweeney, who represented Ken Abraham in this venture, made everything a bit easier.

Most of all, thanks to Elizabeth, who has lived with me, laughed with me, and listened to me. She's supported me all along the way, and together we've climbed many mountains, including this one. She now proudly represents North Carolina in the U.S. Senate, and already has an impressive record. I continue to wonder how she does it all.

Contents

Contents

Italy: April 14, 1945

FRANCE

GERMANY

SLOVAKIA

AUSTRIA

HUNGARY

SWITZERLAND

•Milan

Venice•

Turin•

Po R.

YUGOSLAVIA

Genoa•

Bologna•

Ravenna•

Genghis Khan Line

Castel d'Aiano•

Gothic Line

Pisa•

Leghorn•

Florence•

Arno R.

Tiber R.

Ligurian Sea

ITALY

Corsica

Adriatic Sea

•Rome

Anzio•

Naples•

•Salerno

Sardinia

*Tyrrhenian
Sea*

M e d i t e r r a n e a n   S e a

Sicily

Malta

N

W   E

S

ALGERIA

TUNISIA

0   100 miles

0   100 kilometers

One Soldier's Story

CHAPTER 1

What a Life

He looked so young, just a boy, really, not much more than twenty-one years of age. It wasn't fair that he'd already experienced so much pain and misery in his short lifetime. It wasn't right that his lofty hopes and dreams for the future had been snuffed out by one blast from an enemy explosive device.

But there he was, in the intensive care unit at the Walter Reed Army Medical Center, in Washington, D.C., fighting for his life.

My wife, Elizabeth, and I often visit wounded soldiers at Walter Reed, but this occasion was different. It was Christmas day 2004, and I was about to be discharged from the hospital myself. I had recently undergone surgery in New York, and had been transferred to the medical center in Washington to recuperate.

We were in the dining room shortly before two o'clock, visiting with several young soldiers who had been wounded in the Iraq war, when a mother and daughter spied us. They approached us and introduced themselves as distant relatives of my family. The mother then told us about her son, Craig Nelson, the young man in whose room I now stood. My friend Dr. Charles "Chuck" Peck had informed me of Craig's presence in the hospital, and I had hoped to see him before I left, so the encounter seemed almost providential.

Craig had been badly wounded while on patrol in Iraq a week or so before Christmas. He suffered severe damage to his C-1 vertebra and was paralyzed from his neck down. Now lying in an intensive care unit at Walter Reed, he couldn't move a muscle. He was hooked up to all sorts of medical machines, with various tubes running to his body, an electrocardiogram monitoring his heart, a respirator helping him to breathe, and a tracheotomy in his throat.

Nevertheless, the young man's eyes brightened as I stepped up to his bedside. His mother introduced us: "Craig, this is Bob Dole." Craig's sister joined us around the bed. Craig couldn't speak, but he could hear me and seemed to respond with his eyes.

Looking at Craig, I felt a wave of emotion sweep over me, nearly overwhelming me. It was like seeing a mirror image of myself sixty years earlier. He was tall and muscular, about six feet, one and a half inches, and about 185 pounds, almost identical to my World War II height and weight. For a moment I was back there, in a similar hospital bed, encased in plaster, unable to move, paralyzed from the neck down.

I just stood there at Craig's bedside. I could feel my heart thumping loudly in my chest, my emotions rushing to the surface. I knew the tough road Craig had before him—and his condition was far worse than mine had been.

I reached out my hand—my left hand—touched the soldier's arm, and said, "Good luck, Craig. You're in a great hospital. They'll take good care of you." We stayed only about five minutes.

I looked the young man in the eyes one more time, then turned to his mother, put my arm around her shoulder, and said, "We'll pray for Craig's recovery. Please let me know if I can help."

Unfortunately, a few days later Craig Nelson, another American hero, passed away. I grieved for that family and became more determined that this book would do something to help others understand their pain—and the trauma that so many others have endured because of war.

I've seen these kids in the hospitals and out, people who face seemingly impossible challenges, and I've seen myself in them. Whatever reassurance, hope, and inspiration I can offer them comes out of my own life experiences.

It's said often that my generation is the greatest generation. That's not a title we claimed for ourselves. Truth be told, we were ordinary Americans fated to confront extraordinary tests. Every generation of young men and women who dare to face the realities of war—fighting for freedom, defending our country, with a willingness to lay their lives on the line—is the greatest generation.

In the end, what gets people through a physical or emotional crisis is not new technology or medication. Those things can help, of course. But it's faith that gives you the strength to endure—faith that won't allow you to give up; faith that manifests itself in a ferocious determination to take the next step—the one that everyone else says is impossible.

Adversity can be a harsh teacher. But its lessons often define our lives. As much as we may wish that we could go back and relive them, do things differently, make better, wiser decisions, we can't change history. War is like that. You can rewrite it, attempt to infuse it with your own personal opinions, twist or spin it to make it more palatable, but eventually the truth will come out. Those pivotal moments remain indelibly impressed in your heart and mind. For me, the defining period in my life was not running for the highest office in the land. It started years earlier, in a foreign country, where hardly anyone knew my name.

Dear Mom and Dad,
   What a life! I can hardly believe that I'm living in such a wonderful place. My rest is about over, but I've really enjoyed myself so far. I'm going on a tour this afternoon, also one tomorrow morning. I should see about everything when I'm finished.
   The radio is playing. It reminds me of the times that I've been home playing Norma Jean's records. So far I haven't heard any records by Frank

Sinatra. I guess he isn't too popular over here.

The war news really sounds good. I guess Russia plans on helping us with Japan. Keep your eyes on the news for big things to happen.

Had a fine breakfast this morning, scrambled eggs, bacon, tomato juice, toast and coffee. I sure miss my quart of milk per day. Tell Aunt Mildred to be sure to save some for Kenny and me when we get home.

I ran into a Lt. in Eugene's camp only yesterday but still haven't seen Eugene.

So bye for now
Love
Bob

I folded the handwritten letter dated April 7, 1945—one of the last that I would ever write with my own right hand—and slipped the paper inside an envelope. I had been enjoying a short leave in Rome, but the "wonderful place" in which I was living was a small pup tent pitched in war-torn Italy, near the Po Valley, not far from the mountain town of Castel d'Aiano. I was a

second lieutenant in the infantry "resting" with
the U.S. Army's 10th Mountain Division, an
elite division of soldiers that, just prior to my arrival, had fought one of the bloodiest battles of
the war. Because we were now in a rest mode,
holding the territory that had been won at such
a high price little more than a month earlier, I
had plenty of time to write letters and to think.

The war had finally tilted in our direction.
Since D-day, in June 1944, the Allies had
steadily fought from the beaches of Normandy
toward Hitler's heartland. Although the casualties had been horrendous, the war was winding
down. The Germans were retreating, Auschwitz
had been liberated, the Americans and Russians
were advancing toward Berlin, and it was time
to start planning for the future.

I couldn't wait to get back home—home, to
Russell, Kansas—where I'd been born and
raised; back to Kansas University, where, prior
to the Japanese attack on Pearl Harbor, I had
been studying to become a doctor. I was looking
forward to reestablishing myself as a student-
athlete at KU—well, at least as an athlete. Although I made passing grades, academics had
not been my strength in school. Sports had been
my passion. At KU, I had competed in football,

basketball, and track, and since I'd kept my six-foot-two-inch, 194-pound frame in good shape during the war, I was hoping that I could pick up where I'd left off and make the starting lineup.

Ah, yes, back **home.** A pang shot through my heart when I thought about the young KU woman whom I had hoped to marry . . . until the war separated us for prolonged seasons and our only contact came through letters. I guess she grew tired of waiting for me; while I was still in training, she married an officer in the navy.

Maybe someday I'd fall in love and get married. But for now, like most soldiers who had been away from their families, I longed for the simple things: a home-cooked meal, a comfortable bed, a bath or shower with unlimited hot water—normalcy. Many nights, as I lay in my bedroll and dreamed of how it would be when I got back home, the crackle of distant enemy machine gun fire or the ominous **brruupp!** of a German "burp gun" would jar me back to reality. I was in Italy. This was war.

President Franklin Delano Roosevelt, Winston Churchill, and Josef Stalin had described the Italian front as the "soft underbelly" of the opposition; renowned war reporter Ernie Pyle referred to Italy as "the forgotten front." But apparently nobody had informed the Nazis of these dubious distinctions. In fact, Hitler apparently regarded the rugged Italian terrain as his last hope of fending off the Allied advances; he had personally ordered that every inch of northern Italy be defended to the last drop of German blood.

And a lot of blood was spilled on both sides. The American forces had pushed the Germans back from North Africa and into the "boot" of Italy. Allied victories at Anzio, Salerno, and Monte Cassino had been hard fought, incurring heavy casualties. Meanwhile, Mussolini had been booted out of power, only to be reinstated by Hitler, to no avail. By the summer of 1944, the Germans had grudgingly given up Rome and Florence, and had regrouped to the north in the Apennine Mountains, digging in among the steep mountain ridges. With the Nazis controlling these heights, the advancing Allied forces skidded to a standstill at the German "Gothic Line," a 120-mile heavily fortified barrier stretching across the Apennines north of Florence and blocking entrance to the Po Valley. Beyond that was another line of German defenses known as the "Genghis Khan Line," entrenched in a second chain of mountain ridges. For the next three hundred days the Germans held on tenaciously, knowing that if they lost this ground, the Allies would pour through the Po Valley and make for the suddenly vulnerable southern borders of the Reich, and from there, on to total victory. Despite diminishing air support, the Germans were well dug into their bunkers around Castel d'Aiano, refusing to budge. They fought as men with no hope— either ahead of them or behind them. They were a formidable foe, because they had nowhere else to go.

One of the areas most fiercely defended by the Germans was Mount Belvedere, whose 3,500-foot elevation was considered inaccessible without one's being spotted by snipers. Because they also held the 1,500-foot-high, sheer-cliff elevation known as Riva Ridge on the opposite side of the valley, the Germans had clear sight of any American forces attempting to mount an assault against them from any direction in the punch bowl below. Moreover, by controlling both high observation points, they created a bottleneck through which any Allied troops would have to pass if they hoped to move north from Florence to the ancient university town of

Bologna, and from there, onward toward Hitler's homeland. Three times the Allied forces pushed north against Mount Belvedere, and three times German defenders pounded the attackers back, with heavy loss of life.

Finally, on February 18, 1945, a few weeks before I arrived in the region, the 10th Mountain Division, the last American army division to enter the fighting in Europe, successfully ascended the sheer cliffs of Riva Ridge in the dark and launched an assault on the tenaciously held German positions. Under the cagey leadership of their commander, General George P. Hays, the men of 10th Mountain caught the entrenched Germans by surprise, dislodged them from the mountain, and then withstood seven ferocious counterattacks by the enemy—all in one night. By daybreak Riva Ridge belonged to the Allies. They then crossed the valley and scaled the heights of Mount Belvedere, which had been laced with German land mines and strung with clear trip wires about chest high. Attempting to keep their footing while climbing the mountain in the dark, more than a few soldiers stepped gingerly around a land mine only to bump into a chest-high line that set off a killer blast.

So confident were the Germans that they literally stood at their observation points and watched Hays's men coming. Inch by inch, climbing, crawling on their bellies, slipping and sliding, the 10th Mountain Division would not be denied. Through the smoke, dirt, and fog, the Allied troops engaged the Germans one by one, doing what the Nazis thought had been impossible—driving them off the mountain—and every piece of ground the 10th took, it kept.

The "elite" 10th Mountain Division, as the Germans came to call them, was composed of an unusual assortment of soldiers, made up of European and American ski champions, Olympic swimmers, rock climbers, and Ivy Leaguers, from universities such as Harvard, Dartmouth, and Yale. Hundreds of men in the 10th Mountain Division had earned doctorates from prestigious universities.

But the 10th Mountain was anything but a bunch of rich softies trying to ski their way through World War II. Quite the contrary: its ranks contained some of the toughest, best trained, most highly conditioned fighting men in the world.

The idea for a division of soldiers specially trained to operate in the snowy, cold, rugged mountains originated in the heart and mind of Charles Minot "Minnie" Dole—no relation—an American ski enthusiast inspired by the prolonged resistance of Finnish troops fighting against the Soviets in a 1939 border war. The heroic Finns had held out largely due to their ability to traverse the snow-covered landscape on skis.

Since the outbreak of war in Europe, Charles Dole had relentlessly lobbied Roosevelt's war offices to create a similar force skilled in mountain fighting, with the ability to operate on skis. He wasn't alone. The notion was not far from many military minds that the Japanese might attack the United States in Alaska, or that the Germans might attempt to encroach upon our mainland, invading through the snowy, mountainous terrain of Canada. This wasn't as far fetched as it may sound. The Japanese had already proven their audacity by attacking Pearl Harbor, and nightmares of the Nazis launching a similar surprise attack across the borders of Maine, Michigan, or North Dakota seemed all too plausible.

Finally, in 1943, Roosevelt concurred with General George Marshall on the usefulness of such a division, and a call went out for volunteers. World-class skiers, forest rangers, trappers, mountain mule trainers, cowboys, and woodsmen responded, and soon all were training on skis in the mountains surrounding Camp Hale, Colorado. Besides their ability to ski around rocks and trees while carrying heavy backpacks and M-1 rifles, the men of 10th Mountain became adept in silently scaling rock cliffs using ropes and hooks. What many critics regarded as a wartime country club soon matured into one of the crack divisions in the United States military.

Fresh out of Officer Candidate School at Fort Benning, Georgia, I shipped out for an unknown destination in December 1944. I had enlisted in 1942, was called up in 1943, and by early 1945, I was assigned to a camp of replacement officers outside Rome, Italy. Second lieutenants were a vanishing breed due to the devastating toll the war had taken on our troops' platoon leaders. Many young officers had been gunned down by German artillery during the Italian invasion. One group hit particularly hard was the 10th Mountain Division, which had lost many valiant warriors in the battles to secure the high ground of Mount Belvedere. That's where the army sent me. I thought it mighty odd that a kid from Kansas who had seen a mountain up close only once in his life would be assigned to lead a platoon of mountain troops. We Kansans didn't ski much. But the army didn't ask my opinion. Nor had it bothered to inform me of the severe losses taken by the company to which I had been assigned, the 85th Regiment, 3rd Battalion, I Company. The company commander, Butch Luther— an All-American football player from Nebraska—and half the lieutenants had been killed during the hellacious attack on Mount Belvedere. No wonder the forty or so men of the 2nd Platoon didn't go out of their way to get to know me when I arrived. They figured I wouldn't be around long.

CHAPTER 2

Fragile Flesh

I was easy to spot as the new guy. I was the one with a clean uniform. Fresh faced, wide-eyed, and twenty-one years of age, I wore a tank jacket and tucked my pants neatly into my boots. Even though I'd grown up in a poor family, my mom had always been fastidious about her kids' appearance. Not many men in town went to work each morning with pressed creases in their overalls. My dad did. In Italy, I was determined to keep up the platoon's morale, to demonstrate that even though we were living in bleak conditions, we didn't need to denigrate ourselves further by allowing our spirits to spiral downward. So I shaved every day possible and kept my clothes neat and my kit clean. **Mom would be proud of me**.

The guys in the platoon eyed me coolly at

first. As I said, they'd seen second lieutenants come and go; they weren't about to establish any deep bonds with me. Nevertheless, I introduced myself to each man. "I'm Lieutenant Dole. I'm going to be leading the platoon." In my peripheral vision, I caught a few guys rolling their eyes, as if to say, "Yeah, right." Some nodded, some saluted casually, some shook my hand. Most of the men simply stared back at me blankly. They knew, even if I didn't, that I was a marked man.

One of the first guys I met when I took over the platoon was Sergeant Frank Carafa, a stout, strong, dark-haired, dark-eyed man who had been in the army prior to the outbreak of the war. The sergeant had been acting as platoon leader since the platoon lost its lieutenant in battle. Carafa's eyes bore into me as he presented me with the dead lieutenant's binoculars, maps, compass, .45-caliber pistol, and other materials. It was no secret that many battle-weary guys on the front held the hastily trained officers from Fort Benning—"Ninety-Day Wonders" as we were known—in lower esteem than the officers who had worked their way up through the ranks. As a second lieutenant in the infantry, I had no delusions of grandeur, though. I had heard the stories of how the German snipers particularly went after the guys with stripes on the backs of their helmets and the soldiers with the binoculars and map cases, knowing that if they could pick off the officers and the radio men, they would disrupt the chain of command. Our guys understood that.

I nodded at Sergeant Carafa periodically as he went through his "orientation" talk, explaining to me how the platoon had been operating. He was a good, seasoned soldier, one who had been serving his men well. I had no desire to usurp his position in the platoon, even though, technically, I was now the guy in charge. I had a strip of brass on my collar, but this man was golden.

"How long have you been running the platoon, Sergeant?"

"Since Belvedere, sir." He didn't need to explain.

I turned to him and said, "All right, Sergeant. There won't be any changes. We'll run the platoon as you have been, until I get the knack of it."

The sergeant looked back at me in surprise. "Yes, sir," he said.

I attempted to be equally low-key with the other guys. Since they had just finished a major combat push, and were awaiting supplies and replacements, things were relatively slow around the camp during my first few weeks in command. The lull in the action gave me an opportunity to get acquainted with the men, joke around with them a bit, and establish some rapport. I purposely tried to stay out of the command tent as much as possible, eating with the guys, sitting out on a log or on the ground, talking casually with my new comrades, allowing them to get to know me, and learning something about their capabilities. If I'd learned anything from athletics in high school and college, and my training in the army, we were going to need one another, to defend one another, to depend on one another. They say that every soldier on the front saves every other soldier's life, and has his life saved by another soldier nearby.

I could not imagine how much I would need these men soon.

Part of my duties as a second lieutenant involved leading patrols out at night, to watch for any enemy incursions into our lines. There was always movement once the sun went down, by our guys and by theirs. It was like a shooting gallery in the dark. Bullets whizzed by in the trees; it was impossible to see where the stuff was coming from . . . until it was too late. The snap of a twig or the slip of a foot on a rock could spell doom for the entire patrol. And then there were the land mines, which were strewn all over Italy, it seemed.

I listened to the noncommissioned officers and learned from them. "It would be better if you stayed in the center of the patrol," they said. "That's what most of the lieutenants do."

"No, I'll take the point." I didn't know much about leadership at the time, but instinct told me that a leader had to be out front. He must be willing to endure the fire, rather than hide from it or let somebody else take the brunt of the blows. I wasn't looking to be a hero. In fact, I didn't even want to be there, but now that I was, I was determined to do a good job.

During one of my first night patrols, March 18, 1945, we encountered the enemy in the dark. Coming under fire, I and others pulled grenade pins and threw our grenades in the direction where we thought the Germans were hiding. I waited and watched to see where the grenades would explode. Too late, I discovered that one of the grenades must have hit a tree or something and had rolled back in my direction. Just then, I saw a blast and felt a searing-hot pain in my leg, ripping my flesh open. It was just a surface wound from the grenade fragments, but it hurt like crazy. We kept pushing on in the darkness to no avail, hoping to capture the Germans and extract some information about their comrades who were dug in to pillbox-type bunkers connected by tunnels in the hills all around us. From those same bunkers, the Germans blasted away at us night and day. Their .88-millimeter guns blew gaping holes wherever the guns' shells landed.

Following my first misadventure on patrol, the medics patched me up as good as new. I wrote home, but carefully avoided mentioning the wounding. No use worrying Mom and Dad. Instead, I tried to keep the letter on an upbeat tone. On March 31, 1945, writing on red American Red Cross stationery, I scrawled:

Dear Mom and Dad,
Well, I haven't written for about a week so I imagine that you'll be pretty anxious to hear from me.
Haven't been doing much the past couple of weeks; we're resting now,

but not for long. I've been hoping that I might get a pass but so far I've been unfortunate. I'm a new officer in the company so consequently I've been catching all the assignments. Eventually, I'll get caught up with the other officers and maybe I'll get a few of the breaks.

I really came close to seeing Eugene a few days ago. I found his company and it was just my luck that he was on pass to Rome. He is only about an hour's ride from me so I should see him when I get back. I told some of his men where he could find me so maybe he'll look me up if I can't get time to see him.

I've gotten 18 letters in the last two days, most of them from Norma Jean and "Mom." Also had one from Kenny, "Pie," "Bud," Mrs. Ruppenthal, and Angeline. One of them was dated 23 March which means that it took only seven days. Pretty good service, in fact the best so far.

Keep the packages coming. I received one 3 days ago; it contained candy bars and sunflower seeds.

I'm the "OD" [officer on duty] today, there's not much to it, but I'm not getting much sleep nevertheless. I think I get more sleep on the lines than I do when we are in a rest area. There's really not much danger on the lines, a few artillery and mortar rounds now and then. The thing I dislike most is night patrols. I think I age about ten years on each patrol. You can't see a thing, which is probably the worst part of it. I've led quite a few patrols, and I'm gradually getting used to it.

Tomorrow I intend on reading all my letters again and I'll try to answer some of the questions in them. I'm going to sleep for a while now, so don't forget the packages.

I hope Russell Hi won the State Tournament.

So long for now
    Bob

The break in action gave me another chance to get to know some of the guys in the camp a little better—not well, but better anyhow. There was Corporal Anthony Sileo and Sergeant Devereaux Jennings, an Olympic skier. Staff Sergeant Aulis "Ollie" Manninen hoped to compete as a cross-country runner in the 1948 Olympics, if indeed the games were to be played; and then there was Technical Sergeant Stan Kuschik, Privates Fred Romberg, Arthur McBryar, Eddie Sims, my radioman, and Sergeant Frank Carafa. There were others, of course—quiet heroes the world will never hear about—all good men, all longing to be home.

As the temperature climbed, and the mountain snows melted in the spring sunshine, we knew the command would soon come down ordering us to conduct a major push over the Italian ridges toward the Po Valley. The plan was to cut Hitler's troops off so they could not offer any reinforcements to bolster the Nazi resistance against the two-pronged thrust of the Americans and the Russians heading toward Berlin. The 85th Regiment intended to launch its portion of the spring offensive on April 12, 1945, in an area known as Monte della Spe. What the Army brass called "Operation Craftsman" was labeled in somewhat more colorful language by the guys in the infantry.

Perhaps sensing our impending attack, the Germans increased the bombardment from their artillery placements in the hills above us, forcing us to remain ensconced in foxholes nearly twenty-four hours a day, waiting for our orders, trying to keep our heads down, and trying to suppress the urge to urinate in our own pants. Were we afraid? You bet we were.

We had good reason to fear what was coming up. The German soldiers who controlled northern Italy knew every inch of that ground, and as the American forces moved from one area to the next, they found enemy mortars aimed in their direction. Day after day, the .88-millimeter guns blasted away at us. Sometimes the artillery pounding continued all night long, making it all but impossible to get a good night's sleep.

Guys dug in, two to a foxhole usually, with one man keeping watch while the other tried to catch some shut-eye. It doesn't sound all that uncomfortable until you imagine lying all day and night on your belly on the hard, cold ground . . . or worse yet, with the onset of the spring rains, in the mud. With the rains came an enveloping, misty fog, so much so that between

the smoke from the German barrage and the
gray haze surrounding us, we had to strain to see
the guys over in the next foxhole, just six feet
away. We didn't dare light a campfire—we knew
the enemy was out there somewhere nearby; no
use giving them a free shot—so we couldn't
cook anything. What food we ate came from our
own kits, usually cans of beef stew or some other
vaguely edible concoction. Days were inter-
changeable; nights seemed interminable.

Finally, the order we'd all been expecting
came down from the generals. The 225,000-
man 5th Army would mount a massive assault
that would, they hoped, push all the way to the
Po River Valley, and beyond, accelerating the
presumed Nazi collapse, the surrender of Berlin,
and the end of the war in Europe. The men of I
Company of the 10th Mountain Division—my
guys and I—waited for the word to move out.

The company commander, Captain Jerry
Bucher, met me between artillery blasts and
gave me my instructions: "The breakout will begin 12 April, at 0800 hours." The plan was for
the bombers to come through first, to give the
Germans a wake-up call they'd never forget.
Then the full fury of our artillery power would
be unleashed on them. Finally, if there was any-

one left on those hills out in front of us, we infantry guys would go up and flush them out. It seemed like a good plan, but on April 12, the morning skies turned into gray soup. You could practically bottle the mist in the air.

Bad news. The bombers wouldn't be coming . . . at least not today. They were grounded by the fog socking in the airstrip at Pisa.

We hunkered down in the foxholes again and waited.

Worse news. On the evening of April 12, 1945, we received word that President Franklin Delano Roosevelt had died. At first we thought it was just a nasty rumor, perhaps nothing more than a bit of misinformation foisted upon some unsuspecting soldier by the German propaganda machine. But before long, the rumor was confirmed. Roosevelt was dead.

For many of us, it was hard to recall a time when FDR had not been president. Having pulled, pushed, and prodded our nation out of the Great Depression following the collapse of the stock market in 1929, Roosevelt had again rallied Americans following the blatant Japanese attack on Pearl Harbor, on December 7, 1941. His voice had become familiar through his radio messages laced with hope in the face of stark reality. Now our commander in chief . . . was dead.

Throughout most of his adult life, although physically crippled by polio, FDR had exuded confidence and vitality. He raised our spirits and marshaled our energies as he led the United States through some of our darkest hours. He carefully avoided being photographed in his wheelchair, but I really don't think his disability mattered to anyone but him. The nation trusted him, responded to his leadership, and elected him to the presidency four times—more than any other U.S. president in history.

Untold numbers of American families had a painting or a photograph of President Roosevelt in their living rooms, offices, or some other central location. To put that in perspective, think about how many private homes you've been in lately that have had a picture of a current or recent president on the living room wall. If you live outside the Washington Beltway, you can probably count them on one hand.

By the time I arrived in Italy, Roosevelt was an American icon; yet he was not immortal. In March 1945, his deteriorating health seemed exacerbated by the war and the burdens of his office. With a plan of action firmly in place to bring the war to an end, and victory in Europe nearly in sight, the president decided to leave Washington for a brief vacation at the "Little White House," in Warm Springs, Georgia.

It was there, on April 12, as his likeness was being captured on canvas, that the president flinched in his chair. As he instinctively clutched his head with his left hand, an assistant rushed to him. "Do you need help, Mr. President?"

"I have a terrific headache," Roosevelt said.

Then he collapsed and lost consciousness. He clung to life for slightly more than two hours, while an attending physician tried everything he knew, including a shot of adrenaline to the heart, to revive the dying man. At 3:35 P.M., on April 12, 1945, the leader of the free world slipped into eternity. Before nightfall, Vice President Harry Truman was sworn into office as the thirty-third president of the United States. I've always wondered how the world might have been different, and what might have happened to me, had President Roosevelt lived a few more weeks.

In Berlin, Hitler learned of Roosevelt's untimely demise, and took it as an omen that the war's momentum was about to reverse, in favor of the Third Reich. He was wrong. Within a few weeks, Hitler would commit suicide in his bunker, and the Third Reich would be decapitated. But, of course, we didn't know that then.

A lot of the soldiers in the Italian trenches took Roosevelt's death hard, me included. Some lay in their foxholes and wept. The news passed swiftly from foxhole to foxhole, followed by sobs, curses, and other outbursts. No one was unaffected.

Not that the president's death would change anything. We still had to push through to the Po. Hitler still had to be stopped. The Japanese would have to be vanquished as well.

But Roosevelt's death was one more reminder, on a day when we really needed no more reminders, that we were all fragile flesh, extremely human, and vulnerable. FDR's battles were now over; within a month, our nation's battles in Europe would be over. I had no idea that the battle of my life was about to begin.

CHAPTER 3

Hill 913

Many men in the foxholes thought it was because of the president's death that the bombers didn't show up on April 12. Maybe so, but a far more likely cause was the inclement weather over the Apennines. Heavy fog shrouded the skies, and a wet mist hovered in the air all the way to the ground. By midmorning, we knew our guys weren't coming.

Maybe tomorrow.

Not that the Germans had decided to take a day off. Our bombers might not have been blasting the hillside to bits, but the Nazis continued to pour artillery rounds down from the flat-topped hill in front and above us, blowing huge holes everywhere their shells landed. It was going to be another long day in the foxholes.

The weather was no better on April 13, and

the bombers remained on the ground again. The dawn of April 14 looked little improved, but around eight in the morning the fog began to burn off.

"Get ready," I told my guys, bending low and moving quickly from one foxhole to the next. "It won't be long now."

Within half an hour, we heard the first drone of heavy bombers above us. A few seconds later, some of the most intensive bombing of German positions in the entire Mediterranean theater began, as the bombers pummeled the mountain and hillsides in front of us. For the next forty minutes, wave after wave of bombers dropped their devastating loads on the German bunkers. Fighter bombers streaked lower, turning the hillside into dust fragments as thousands of pounds of high explosives blasted the German positions. Napalm bombs followed, touching off what appeared to be hell on earth. Large chunks of the mountainous terrain burst into flames and remained on fire. Then the medium bombers went in after the German machine gun positions.

The sounds of our planes veering off and returning to base had barely disappeared when our artillery—more than two thousand pieces

strong—opened fire on the Nazis. The sound itself was horrifying. Imagine the noise of a thousand Fourth of July fireworks displays going off
at the same time, and you'd have some idea of
the deafening, booming explosions in our ears.
Even the older guys in the platoon agreed—it
was a bombardment unlike anything any of
them had ever seen before. The barrage continued for somewhere between thirty-five minutes
and an hour; it was hard to tell the difference between smoke plumes and dust clouds hanging in
the air. The sky turned an opaque gray. By ten
o'clock, we could hardly see the area directly in
front of us, a long ravine sloping toward a
steeply graded mound simply identified as Hill
913 on our maps.

The bombardment evoked mixed responses
from the men around me. We were thrilled that
we were getting such advance firepower; surely
that would ease our task of taking the ground on
foot. On the other hand, we couldn't help but
wonder what in the world was **out there**, if the
big guys felt that the German battle-proven
troops—the 334th Grenadier Division, we later
learned—merited such a drubbing just to be
softened up for us. Some guys held their ears,
others covered their heads with their arms. Still

others sat up cockily, grinning openly. "Nobody could survive a pounding like that!" a few guys mouthed, though their words were obliterated by the horrendous noise.

Then, as suddenly as it had started, the bombardment stopped.

The air had now turned an eerie bronze-brown from the dusty grit hanging over the area. "Move out!" came the command. I waved the platoon forward. We scrambled out of the dirt in our foxholes into the dirt in the air, in the general direction of Hill 913. Our orders were to cross the valley and make our way up the hill, secure its summit, and continue on for three miles, mopping up any German resistance as we went.

The plan was for the 85th Regiment to take the left side of the valley, moving down a slope into the ravine then across about a thousand yards of open terrain, a desolate clearing. On the other side of the clearing was a long barrier of thick hedgerows, probably planted a hundred years ago to create a natural boundary between two Italian farms. At several points the hedgerows gave way to waist-high stone walls. They may have been quaintly beautiful at one time, reminiscent of the rolling farmlands of

New England or the stone "slave walls" that still remain in southern portions of the United States as poignant reminders of the War Between the States.

The hedgerows and stone walls fronting Hill 913, however, served simply as obstacles to our getting up the slope to our destination. More ominously, if past experience proved anything, the Germans had probably sown that valley near the hedgerows with land mines and booby traps. It would be almost impossible to get across that field quickly. Most likely, we'd have to traverse those thousand yards on our bellies, crawling on our hands and knees, and praying all the way that a few of us would make it.

We inched our way along silently, single file, into the valley; the 3rd Platoon on our right, out ahead, the 1st Platoon behind us, all of us hoping against hope that the Allied bombardment had taken care of the Nazis.

About the time the platoon in front of us got over a stone wall, and forty feet or so into the clearing, someone stepped on a land mine. Someone else hit another. Then the world exploded. **The Germans were alive!** The massive bombings apparently had shaken their defenses, without destroying them. They started pouring artillery, machine gun, and mortar rounds into the clearing in front of us, mowing down dozens of American soldiers, shredding others, pulverizing still more. A number of boys jumped to their feet, attempting to run to shelter in recently created shell holes. As they did, they were tripped up by wires and met by "Bouncing Betties," grenades attached to the wires that had been strung low to the ground. When the wire was stretched, it bounced the grenade off the ground, often into the face or body of the soldier who had tripped the wire. The grenade exploded, spreading a deadly swath of shrapnel in every direction, wounding or killing even more soldiers.

Guttural grunts, thuds, and screams added to the bedlam. Most of the guys in the 3rd hit the ground, scrambling for any cover they could find. Our guys in the 2nd Platoon did the same. Large mortar rounds continued to riddle the area around us. In the hills behind a large stone farmhouse, we now realized, were embedded the pillbox bunkers out of which the Germans were firing mortars as fast as they could load them. Stuff was coming at us from every direction.

That's when we heard the machine gun fire

coming from a farmhouse on the hill. The German machine gunners were notorious for their barbarism, reportedly firing on the living, the wounded, or the dead with equal viciousness. When they saw a soldier go down, they'd continue firing. And at least twenty-five or thirty of our boys were already down.

Raw anger surged through me as I saw so many of our guys falling. I called for cover artillery fire, and my radioman, Eddie Sims, staying close to me, relayed the message. A smattering of fire erupted near the hedgerows, but the German machine guns kept crackling.

Just then, the company runner dived onto the ground beside me. "The captain wants to see you and Carafa up front," he said, already climbing onto his hands and knees, up to a bent-over position, and running back toward the front line, a short distance ahead, behind a stone wall.

Sergeant Frank Carafa and I made our way to Captain Jerry Bucher, the company commander. Captain Bucher looked at the sergeant and me, and said, "Take the platoon around the left flank, and stop that machine gun fire." When the sergeant and I got back to our guys, we explained what we were going to do. But rather than letting the sergeant take the lead, as he had expected, I waved him off.

"I'll take 'em, Sergeant," I said. "You provide the cover fire." The sergeant looked somewhat surprised—he obviously thought he'd be the guy tapped to step into the murderous open field of fire on the other side of that wall—but this was no time for discussion. Carafa was a good soldier. He did what he was told.

I took part of a squad from the platoon with me toward the left, while Carafa rounded up the rest of the guys and got them in position, ready to fire at the farmhouse straight on, as we attempted to come in from the side. There was a small wooded area beyond the ravine. If we reached it, we could sneak close enough to the farmhouse to use grenades to knock out that machine gun nest and some of the bunkers around the house.

Carafa's guys started firing, and I took the point, leading the men out across the field on our bellies. Spreading out, we inched forward, crawling from shell hole to shell hole, hiding behind any defensive barrier we could find. Meanwhile, the Germans kept hammering us with heavy artillery, burp guns, and the occasional, stuttering **rat-a-tat-tat** of machine gun fire.

Several of our men were hit and slumped over on the ground. Several more fell victim to the land mines.

I had to get to that machine gun.

My heart was beating so loudly I was sure the Germans on the hill could hear it. A few of us reached one of the hedgerows. I took a deep breath and tried to collect my thoughts. We'd found an oasis in a dry, gritty desert, a bit of safety at last. But it was an illusion. Moments later, a heavy machine gun opened fire on us from the farmhouse.

Sims called for mortars. I raised my hand and nodded, and the men and I scrambled out from behind the hedgerow and started up the steep, rocky grade, heading toward the farm-house. The Germans might get some of us, but they weren't going to get us all before one of us got them. We got about fifty or sixty feet away from the hedgerow before the machine guns crackled again. I pulled the pin on a grenade and whipped the grenade toward the sound of the machine gun fire. The grenade exploded with a blast, but short of the house.

Private Fred Romberg, our first scout, had a better angle than I did. He half stood so he could throw better, get some distance on his grenade toss, but he had no sooner risen up when machine gun fire cut him down. He writhed in the air and fell face forward into the dirt. His helmet fell off and rolled in front of him. As in a surreal movie, Romberg's helmet rolled away from its owner and tumbled back down the slope, clanking against a few rocks before it finally came to a stop.

I wanted to run to him, but I knew that if I did, there'd be two of us dead in the dirt. **Romberg, you're a good man. I won't forget you**.

I peered through the silt in the air, trying to spot our second scout—he was out there somewhere—when a huge blast hit nearby. I dived into a crater hoping for cover as more shells hurtled through the air and exploded beyond us. Still on my stomach, I crawled out of the shell hole. Remnants of my men lay all around me. Some were dead; some were wounded.

Sims. Where's Sims? Where's my radioman? We've gotta get help. I gotta get my guys outta here. They're getting chewed up out here.

Then I saw him. He was slumped in a bloody heap on the ground, still clutching the radio. I couldn't tell whether he was dead or alive.

I scrambled toward him on all fours, the

shards of metal from the shrapnel pricking my hands and knees. **Keep moving. Don't stop. He's right there. You can get him**.

It seemed to take forever to get to him. When I reached him, I turned my back at an angle to the farmhouse, grabbed Sims by his shirt, and started dragging my radioman back toward a shell hole and, I hoped, some measure of safety. Sims wasn't moving; I wasn't sure he was even breathing, but I was trying my best to bring him back.

Too late. I felt a sting, as something hot, something terribly powerful crashed into my upper back behind my right shoulder. I'd often been hit in the back by a hard-charging linebacker when I'd caught a football in high school or college. Occasionally, I'd been thrown to the ground and stomped on by a couple of college linemen after making a catch. But nothing in my life ever hurt like the shock that seared through my shoulder just then.

My body responded before my brain had time to process what was happening. As the mortar round, exploding shell, or machine gun blast—whatever it was, I'll never know—ripped into my body, I recoiled, lifted off the ground a bit, twisted in the air, and fell face down in the dirt.

For a long moment, I didn't know if I was

dead or alive. I sensed the dirt in my mouth more than I tasted it. I wanted to get up, to lift my face off the ground, to spit the dirt and blood out of my mouth, but I couldn't move. I lay facedown in the dirt, unable to feel my arms. Then the horror hit me—**I can't feel anything below my neck!** I didn't know it at the time, but whatever it was that hit me had ripped apart my shoulder, breaking my collarbone and my right arm, smashing down into my vertebrae, and damaging my spinal cord.

At first, I couldn't feel anything. I had no sensation of my limbs, and I couldn't move my face to see them. Had I been able to move, I would have seen my arms splayed out in front of me above my head. Inside the remnants of my frayed, blood-covered coat sleeve, the right arm was dangling from my body. I had already gone into shock.

What happened next I can only tell you with the help of the guys in the platoon who over the years have filled in the many blank spots in my recollection of those moments. One of those guys was future Olympian Ollie Manninen. While running forward toward the hill, Staff Sergeant Manninen saw that I had been hit and was stretched out facedown on the ground with blood all over me. He crouched down to see if I was still alive, then pulled me behind a section of the wall, out of the direct line of fire. Ollie's quick thinking and courageous action probably prevented me from being shredded by the German machine gunners.

Then Ollie did as he had been commanded. He left me lying there all alone and kept moving ahead toward the deadly farmhouse. The medics would be along soon, he presumed, and unless someone could knock out that machine gun, it wouldn't matter where he left me lying.

My body was numb, but my brain was still active, although blurred. I recalled that Carafa and his guys had been giving us cover fire from somewhere to our right. Maybe he or someone else was still there.

"Sergeant Carafa," I moaned. **Sergeant Ca . . . raaaa . . . fa,** I called with all the energy I could muster.

One of the guys pointed me out to Carafa. "That's the lieutenant! He's calling for you."

Frank Carafa later recalled vividly the mixed feelings he had about trying to save me that morning:

I don't know how much time elapsed, but all of a sudden, I heard him call my name. "Sergeant Carafa. Sergeant Carafa!"

I just ignored him. I realized he was hit, and I was just too scared to go out there. He kept calling me, and finally my men started saying, "Hey, Sarge. The lieutenant's been hit and he's calling for you."

What do you want me to do? I said to myself.

But I had to do something, because if I didn't, the men would lose respect for me. So I crawled down close to the opening of the ravine, and saw the whole squad of men on the ground. I didn't know at the time how many had been shot. I didn't know what to do. I just lay there crying on the ground. Then I saw one man move. Another one, about ten feet away from me, moaned. Seeing him, I don't know, something came over me, and I started crawling out to see if I could help him. When I got to him, I started dragging him back.

Not realizing what I was doing, I left him with a couple of other soldiers and I turned around and started crawling back out. The next two guys I came to were already dead.

I crawled on to the next person. He was wounded, so I brought him back. I just turned around and started crawling out there again. I could hear the bullets going over my head. I kept crawling. Two more were dead.

Sergeant Carafa crawled up along a gully while machine gun fire closed in on him. He later said,

I was so scared I wet my pants. I went by instinct. I don't know where I was getting the strength. It had to be God helping me. I was praying; I was crying. I finally reached the lieutenant, about sixty or seventy yards away.

By now, I was barely conscious, still stretched out on the ground and bleeding profusely. Carafa continued,

I tried to move him. His arm was outstretched, so I grabbed his arm and I yanked him. He gave a holler and passed out. I couldn't budge him. I was 5'5", 145 pounds. He was a six-footer and close to two hundred pounds. I just kept praying to God to help me. I finally got some strength and I started dragging him. When I couldn't drag him anymore, I just shoved and rolled him along, down the incline. I was exhausted. I was praying, "God, help me! God help **him**!" He was all shot up. His whole right side, from his shoulder down to his waist. But he was still alive. Face white, and barely breathing, but alive.

Fading in and out of consciousness as I was, I don't remember much of that, but I do recall the searing, excruciating pain that wracked my body when Sergeant Carafa attempted to yank me by my arm. It felt as though what remained of my arm was going to come off in his grip. I conked out after that, although I was later vaguely aware of being rolled over and over again, my head and face bouncing off the rocky terrain like a softball hitting the pavement.

Technical Sergeant Stan Kuschik, second in command in the platoon, came alongside Sergeant Carafa. "Do you have any morphine?" Carafa asked.

"Yeah, I do." Morphine was not readily issued to infantrymen, only to medics, but Stan

Kuschik had picked some up from a fallen medic. The Nazis gave little regard to the Geneva Convention rules of engagement that encouraged leniency toward medical assistants. The Nazis would shoot a medic as quickly as anyone else in their way. Indeed someone had radioed for the medics, and two of them got killed trying to get to me. It was unlikely that any more medics would be along today, not until it was much safer. Kuschik and Carafa realized that might be too late for me.

"Better give the lieutenant a shot," Carafa said to Kuschik. "He's in pretty bad shape."

Kuschik agreed. He later recalled, "The lieutenant was gray, the way they got before they died." The sergeants could see down into the gaping hole in my jacket, into my body, where my shoulder had once been. Kuschik bandaged me up, gave me a shot of the morphine, then dipped his finger in my blood and painted an "M" on my forehead. It was an old battlefield precaution. If help did happen along, when they saw the "M," they'd know that I'd already received one shot of morphine to help numb the pain. A second shot of morphine, if administered too soon, could be a fatal overdose.

Since I was now lying on my back, Carafa

placed my arms over my chest. The battle was still raging all around us, artillery fire was still coming in, and even in my dazed condition, I knew that Carafa and Kuschik had to move on. The prebattle orders had been extremely clear: leave any wounded to await the medics who would be coming behind. It was a tough but practical order, especially in a battle where many casualties were incurred. If everyone who came across a wounded soldier stopped fighting and stayed with the wounded, we'd never take Hill 913.

I lay on the ground looking up at Kuschik, a big man with a heart to match. I tried to speak to him; nothing came out but gurgles. Kuschik looked down at me with compassion, and made a decision to disobey an order. He called to Arthur McBryar, a tall soldier from Tennessee. McBryar had been slightly wounded in the leg, and wasn't able to move quickly. "Stay with the lieutenant until the medics get here," Kuschik told him. Then Sergeant Kuschik rejoined the men still fighting their way up the hill toward the farmhouse.

McBryar remained with me, nervously trying to carry on a one-way conversation amidst the roar of the battlefield, hoping that by keeping my mind occupied he could help me hold on till help arrived. I was conscious, but only my eyes could move. I couldn't even unclench my teeth. At one point, McBryar pressed a bandage into my wounds to try to slow the bleeding. By now, my blood had soaked my jacket and uniform and was dampening the ground beneath me.

I finally mustered the strength to say through my clenched teeth, "How bad is it?"

McBryar pulled the bandage away and looked inside me. He patted the bandage back into position. "You're gonna be fine, Lieutenant," he lied.

For the next six hours, McBryar watched for anyone who looked like a medic, while I lay on the cold, moist ground. I lay there all day long, wondering during those moments I was conscious, if anyone, friend or foe, would ever find us. After several hours, McBryar ventured out to a nearby knoll, hoping to spot a medic. He didn't, but a Nazi artillery blast blew him off his feet. He wasn't hurt seriously, but he suffered a concussion. He stumbled back to near where I was lying on the ground. He didn't talk much after that. Occasionally, I could feel the rain spattering off my face. Or maybe it was simply more of the misty fog. I couldn't tell. My own mind was drifting in and out of a fog.

I wanted to wipe the moisture from my eyes, but I couldn't raise my arm from my chest where Sergeant Carafa had placed it. I could still hear the noise of the battle going on around us, but I didn't know where I was or where any of our guys were. Lying there on the cold ground, I wasn't thinking about the future. I wasn't even thinking about survival. I was think-ing of where I had come from. My mind kept going back, back . . . I didn't know where else to go, so I went **home**, home to Kansas.

It was like watching a movie of my entire life . . . I saw myself as a youngster. I saw the town of Russell and the kids I used to play with as a little boy. I saw my parents at the dinner table, our family at Trinity United Methodist Church. **Now I'm running up Maple Street. I always run. I'm running track in school. Hey, Dad, I think we have a good chance of going to State in football this year. I see Spitzy, our little white dog. Spitzy, what are you doing out in this kind of weather? Mom is going to get after you. Who am I to talk? Look at me; I'm soaked. My jacket's filthy** and all torn up. How'd I get this mud all over my clothes? What's that sticky red stuff? Oh, no . . . Mom's gonna have a fit.

Back . . . back . . . forward for a few moments, then . . . back. It's getting dark, I have to go. . . . I really want to go home. . . .

In the movie **The Wizard of Oz**, when a violent tornado hits Kansas, Dorothy is knocked out in a black-and-white world, only to wake up later to a collage of colors. She goes in her imagination to Oz, a colorful, fantastic place.

In my semiconscious state, near a fever dream of delirium, I felt myself drifting back to Kansas, to my childhood and growing-up days, where everything was black and white.

CHAPTER 4

The Russell Years

I've often said that anyone who really wants to understand me has to go back to my hometown of Russell, Kansas—if not literally, then at least emotionally and culturally. Located in central Kansas, about halfway between Missouri and Colorado on what is now Interstate 70, Russell was originally settled by about sixty families, mostly German-Russian immigrants, from Ripon, Wisconsin (which later became the self-professed birthplace of the modern Republican Party). These early prairie pioneers were soon joined by enterprising souls from England, Ireland, and Wales, as well as a number of people from the eastern United States.

Wherever they hailed from, Russell's settlers had a common objective: inexpensive farmland and a better life than the one they were abandoning. The railroads were giving land to settlers who agreed to farm it, so the pioneers journeyed by train, loaded down with all their earthly possessions. When they arrived at a water stop on the Union Pacific Railroad in the basin between the Saline and Smoky Hill rivers, they decided they were home. It was 1871, only a few years since the Civil War had concluded, and only two years since the last Indian raid, known as the Kits Fork Massacre, had taken place nearby. The town was originally founded as Fossil Station, so named because of the many ancient imprints stamped in the native limestone.

If you flew over Russell and the surrounding environs today, you'd think that this was exactly what Katharine Lee Bates had in mind when she wrote "America the Beautiful," for this part of Kansas is defined by "spacious skies" and "amber waves of grain." But it wasn't that way in the nineteeth century. Back then, the future breadbasket was so desolate that most maps referred to the area as the "Great American Desert." The landscape may have been flat, but nature raised high barriers in the form of tornadoes, drought, and plague-like grasshopper invasions. Winter blizzards dumped waist-high snowfalls, whipped by incessant gales into snow drifts higher than first-story windows. I recall the temperature occasionally plunging to thirty degrees below zero—which was really something in our part of the country. During the summers, scorching winds stripped the land of its topsoil and scattered it into the air. Lack of rain was a constant concern for the early Kansan farmers.

Fortunately, a group of Mennonites introduced a foreign crop, Turkey red wheat, to south-central Kansas. The hardy new strain thrived despite the subzero winter temperatures and the parching summer droughts. Wheat became the settlers' life source, blanketing large parcels of land in waves of undulating grain.

No less than the Pilgrims of Plymouth, the people who wrote Fossil Station's charter did so with high hopes and high standards. The town leaders declared straightforwardly that they intended their community to be comprised of inhabitants with "good dispositions and industrious habits . . . but no person of disreputable character or vicious habits shall knowingly be allowed to become a member of, or settle in, said colony, if in the power of the commissioners or their agents to prevent it; nor shall there be any gambling or tippling houses allowed to be established in said settlement, nor any intoxicating liquors sold therein as a beverage." For the most part, the founders got their wishes, although local citizens eventually developed a slightly more tolerant view of "worldly vices."

Later, in 1871, the residents changed the name of their town in honor of a Union soldier, Captain Avra P. Russell, a New York native who had served in Kansas and died in battle. Although Captain Russell never set foot in Fossil Station, to this day his name conjures up images of resilience and strength; of valiant soldiers, indomitable farmers, and devil-may-care oilmen; of cowboys and country fairs and cottonwoods on the far horizon.

Like the pale, chalky limestone used to build everything from houses to fence posts to yard ornaments, the local culture was firmly embedded in rock-solid, time-tested principles of faith in God, love of family, and loyalty to friends. On that foundation, Russell grew to be a quintessential Midwestern community, a picture postcard of rustic values and plainspoken wisdom. It would be hard to imagine a more democratic setting, or one less sympathetic to stuffed shirts, bigots, or blowhards.

Garrison Keillor's colorful descriptions of

Lake Wobegon and the characters who lived in his fictional village, vividly brought to life though his long-running radio show, **A Prairie Home Companion,** could easily have been applied to Russell. Folks called one another by their first names, took an interest in one another's families, laughed together when things were going well, and cried with one another when times were tough—as they often were for many people in Russell, my family included.

Doran Dole, my father, was born in 1900. His family had lived on farms in upstate New York and Ohio prior to moving to Kansas, where they hoped to buy some land and enjoy a fresh start on the frontier. But raising a crop on the windswept plains of Kansas was even more difficult then than it is today, and the Doles ended up being tenants on the land rather than owners. They paid one-third of their crop to the landlord and lived on whatever was left, which often wasn't much.

The world was still engaged in World War I when Doran Dole celebrated his seventeenth birthday, so along with a friend, he dropped out of school, pretended to be eighteen, and, determined to fight for his country, enlisted in the army. Instead of going overseas, the young patriot was posted to Illinois and Texas for the duration of the war, much to his chagrin.

Upon his return home, Doran Dole met Bina (pronounced **Bye-nah**) Talbott, the energetic eldest daughter of twelve children born to Joseph and Elva Talbott, a similarly hardworking couple who lived on a farm about ten miles south of Russell. Doran and Bina fell in love, and planned to marry.

About that same time, with his small savings Doran rented a storefront on Main Street, which he turned into the White Front Café. The restaurant's sparse wood-floored interior had barely room for a lunch counter with several bar stools, plus a couple of white tables. The room's sole decoration was a sign announcing, WE WELCOME YOU, with four or five small American flags flying above it. When Doran and Bina married in 1921, friends made them a special "honeymoon meal," which they ate at the café. The meal lasted as long as their honeymoon, and vice versa. Doran was twenty-one years of age, and Bina was eighteen. Neither possessed a high school diploma, although both were extremely bright and determined to succeed.

Unfortunately for the newlyweds, few folks around Russell then could afford to dine out, so before long, my parents closed the White Front. Besides, the young couple soon needed a more stable income following the birth of their first baby, Gloria.

By now, Russell had grown to more than two thousand residents. Laid out in a grid, the town was divided by the railroad tracks, but little else. To be sure, my family lived on the North Side, the so-called wrong side of the tracks.

I came along two years after Gloria, on July 22, 1923. I was named after my paternal and maternal grandfathers, Robert and Joseph. Talk about togetherness—Mom, Dad, Gloria, and I squeezed into the tiny three-room house where I was born, just north of the Union Pacific railroad tracks. That's three **small** rooms—and one of those rooms was the kitchen.

When my brother Kenny was born the following year, and Norma Jean the next, Mom and Dad decided to move the family to a five-room house on the corner of Eleventh and Maple streets. That was the home in which my family and I lived until I was nine, when we moved to a larger house a little farther up Maple Street.

Russell received a powerful economic infusion the same year I was born. Less than twenty miles to the northwest, near the small town of Fairport, dreams came true for a group of seven investors when they struck oil on Thanksgiving Day. Money began to flow into Russell as never before, and never since. Along with the sudden prosperity came an influx of new faces, as Russell experienced a growth spurt financed largely by outsiders. The oil boom turned the town into a true American melting pot, although some of the newcomers came with lifestyles that were quite contrary to those of the conservative locals. Before long, the outlying areas along Highway 40 looked like a mini Las Vegas, replete with nightclubs, honkytonks, slot machines, dice tables, and bootleg liquor. Prohibition may have been on the law books, but it didn't prohibit very much—least of all the sale, purchase, or consumption of alcoholic beverages.

Russell's Main Street was spruced up, too, thanks to the gusher of oil money. Existing downtown businesses flourished, and several new ones got off to good starts. For a while, the oil boom brought unprecedented prosperity to many people in Russell. My folks, however, weren't among those who received a windfall. For them, the years following the discovery of oil nearby were pretty much life as usual.

At least their expectations weren't raised. Good thing, because by the early 1930s, the oil boom had gone bust. By then, Dad had moved on from the café to manage the Fairmont Creamery Company, a one-man operation that bought sour cream, milk, and eggs from area farmers. Toiling from dawn to dusk every day, Dad made about fifteen dollars a week to support his family.

Lugging those one-hundred-pound milk cans two at a time, no man worked harder or complained less. Dad prided himself on keeping the creamery spotless, scrubbing and cleaning the place late into the night. A few years after I was born, my father took a job managing the Norris Grain elevator, a receiving and distribution center for wheat, grain, and corn. All in all, he missed only one day of work in forty years, and felt bad about that.

He did take off work each November 11, Armistice Day, when Europe's exhausted nations finally halted the bloodletting of World War I. Dad loved joining in the celebration, sometimes donning his old army uniform and marching in the parade along with the other members of the American Legion. The procession went all the way up Main Street, now paved with red bricks for about half a mile, running north and south through town. The pageantry culminated in the reading of Lincoln's Gettysburg Address, which never failed to bring tears to the eyes of participants and spectators alike.

Whether working at the creamery or the Norris Grain elevator, or at home around his family, Dad displayed a low-key, sardonic sense of humor. Quick to laugh, he could also crack a joke or slip a wisecrack into the conversation without breaking a smile. To the uninitiated, it was sometimes hard to tell if Doran Dole was serious or not. But those who knew him well caught his wit.

Dad taught us kids that in life "there are doers and there are stewers." Dad was definitely a doer. He didn't think that he'd done anything extraordinary by giving more than was expected for his wages. He didn't jabber about doing what was right; he just did it. He embodied qualities such as honor, honesty, personal responsibility, and a sense of duty more than he talked about them.

By the time he opened the grain elevator for business each morning at seven A.M., the aroma of freshly brewed coffee was already wafting through the air, mixing with the scent of the stored freshly cut grain. Dad liked his morning cup strong enough to make your toes curl. When we visited our relatives, he'd drink their usual brew, then deadpan, "Now, how 'bout some coffee?"

The locals loved my dad—they called him Doley—and they regularly stopped by the grain elevator to tap the large urn of coffee that was always available there. Farmers who came in with their grain stayed to talk over everything from the weather to the chances of the local high school sports teams.

Most everyone in and around Russell regarded Doley as a man of integrity, a volunteer fireman who would risk his life to help save a neighbor's house or barn. He was quick to join in with a sandlot softball game or help out with the local kids' sports teams, and during harvest season, he'd often work till the wee hours of the morning helping his farm customers unload their grain. But he was always right back at work the next morning, the grain elevator swept, scrubbed, and open by seven. When my brother, Kenny, and I were just boys, we'd sometimes race home from school during harvest season, grab a quick bite to eat, then run over to relieve Dad at the grain elevator while he hustled home for a quick lunch break. Most other times, Mom took Dad's lunch to him at the grain elevator.

If Dad was the more laid back of my parents, my mom, Bina Dole, lived life at full open throttle. She talked fast, walked fast, and drove fast. Even if she was just driving up the street a ways, she pushed the gas pedal to the floor. It wasn't merely that she was in a hurry; she was perpetually busy, and always had more things to do than she had time for, more that she wanted to accomplish than the hours in one day allowed.

We may have been poor, but that didn't mean we embraced poverty. Mom made sure we maintained our dignity. She taught her kids to respect one another, to respect ourselves, and to watch out for one another. Cleanliness was next to godliness as far as our mother was concerned, and she was fastidious about each family member's personal appearance. "Nobody's so poor they can't afford soap," she'd say. "You never get a second chance to make a first impression" was her motto.

Our home was always immaculate, decorated with rugged furniture in what today some might call "understatement"; for us, it was all we could afford. But Mom scoured the stores for the best buys on everything from carpets to clothing.

What she couldn't find at a bargain, she made herself on her Singer sewing machine. Our windows, for example, were draped with beautiful sheer curtains, handmade with love by our mother.

Mom made most of our clothes, too, especially my sisters' party dresses and pleated jumpers. She refused to discard any article of clothing that could possibly be used for something else. For instance, she cut down her old winter coats to fit my sisters.

You could eat off the floors in our house; they were that clean. And Mom wanted her family to look sharp, as well. We may have owned only a few changes of clothing, but they were always washed and pressed, even if it meant that Mom had to stay at the ironing board long after we had gone to bed. Dad changed his work clothes every day, too—a rarity in those days. Each night, Mom made sure that he had a fresh, clean shirt to wear to work, along with his washed and pressed bib overalls. Our bedrooms, too, had to be neat, beds made before we left for school each morning, no clutter—not on the floor, the desk, nowhere.

Out in the backyard, Mom raised a vegetable garden, and she expected all the Dole children to help. If one of us balked or complained that we couldn't do something, that the task was too hard or couldn't be done, Mom countered with her life message: **"Can't never could do anything.** So get busy."

Half-hearted obedience was disobedience to Bina Dole. She never did things halfway, and she cringed at the thought that one of her children would give less than one hundred percent at anything. To my mom, it was never good enough to "just get by." If a job was worth doing, it was worth doing well.

When Mom scrubbed the floors indoors, she made all four of us kids sit in chairs with our feet hitched up off the floor until it was dry. To four lively children, it seemed like hours before Mom's floors dried. No doubt, she could have sent us outside, or given us something else to do, but I think she wanted us to learn a little about self-control. If one of us dared step on her clean floor while it was still damp, she was pitiless. "You can go outside and cut your own switch right now!" she'd rail as she mopped the floor again. I'm not sure which was more frightening, the actual swats across the rear with the switch or having to go cut the stick that would inflict the pain.

In describing his boyhood, Herbert Hoover recalled sitting in a Quaker meeting house with his parents and other worshipers, waiting for the "divine light" to move them. Young Hoover didn't dare fidget until the Word came. He didn't remember receiving the divine light; but he never forgot the strictness of his parents, and of that culture.

Similarly, my brother and sisters and I may have occasionally bristled over Mom's perfectionism, but that perfectionism was more than offset by her generosity, and her ability to laugh at anything that hinted at self-importance. Although not as overtly funny as Dad, Mom, too, knew how to deliver a deadpan zinger. Throughout the day, she always had a twinkle in her eye and a smile just aching to break out. Beyond that, Mom was always the first person to lend a helping hand to anyone in need. When a friend or neighbor had a serious illness or a death in the family, Mom was right there with a home-cooked meal for the family. Sacrificing her own comfort or convenience for the sake of others was a quality she demonstrated on a daily basis.

A superb cook, Mom made meals that today's fine restaurants could only envy. And she could stretch a food budget farther than anyone I've ever known. Although our family scrimped for cash, Mom regularly prepared scrumptious feasts for us, including homemade bread, soup, a full-course dinner, and an angel food cake or some other tasty dessert. Cakes, pies, cookies, and occasionally even homemade ice cream topped off our dinner, although Dad reminded us almost every night, "Dessert's under your potatoes." In other words, if you wanted the sweet goodies, you had to eat everything on your plate first. Tablecloths and silverware settings were standard fare in the Dole dining room for every evening meal. And none of us dared show up at the table with unwashed hands, disheveled clothing, uncombed hair, or dirty fingernails.

After dinner, everyone pitched in to clear the table and wash the dishes. Then the kids gathered around the same dining room table to do our homework. Meanwhile, Dad relaxed in his large easy chair to read the **Salina Journal**, a regional newspaper, or listen to the radio. I can still recall crowding around our family Philco, listening to Dad's favorite shows, which soon became my own: **Amos 'n' Andy, Fibber McGee and Molly**, and the like. After that, it was bath time, followed shortly by bedtime. The regimen changed little throughout my early childhood.

Mom and Dad didn't work all the time—just most of the time. But they also managed to get out to a local dance every once in a while, or to an ice cream social, and they enjoyed having folks over to the house to play bridge. They weren't antisocial; they just didn't have any extra money for the frills of life. Instead of fancy vacations, we'd take a short trip over to Grandma and Grandpa Talbott's farm, or visit Grandmother and Grandfather Dole. Dad liked hunting and fishing, and once we actually did take a vacation of sorts. When I was about twelve, Kenny and I went with Dad on a fishing trip to Colorado, where a family friend invited us to stay for free in his cabin. It was the only time prior to World War II that I'd ever been that far away from home.

Like our parents, the Dole kids learned to improvise and make our own entertainment. A special treat was a trip downtown to Russell's Dream Theater. Once a week, the theater manager presented what he called the "owl show," a late-night movie for which admission cost only a dime. Better yet, schoolchildren with good grades and attendance records earned a free pass to a Saturday matinee. To add to the incentive, my aunt Mildred Dole offered me twenty-five cents for every perfect paper I received in spelling. I gladly accepted that challenge, and became a stickler for correct spelling—which I remain to this day—and I got to see some great movies.

Hard as he worked, Dad's paycheck provided barely enough to support our family. To help supplement his income, Mom sold Singer sewing machines door to door. Women in the workplace were rare in those days, especially in the role of traveling salespeople.

Mom didn't care. Her family needed the money, so every morning she'd have Kenny or me lift the heavy machine into the back of our old Chevy, and as soon as we left for school, she'd set off driving the country roads around Russell. She knocked on doors, and won the opportunity to haul the Singer into a home. Then she'd demonstrate how easy it was for anyone to sew. Using whatever fabric the homeowner had available, Mom would turn it into something beautiful and useful with the help of the Singer.

On more than a few occasions, Mom would stay up half the night working on a dress or something special for the woman for whom she had demonstrated the machine earlier that day. The next morning, she'd take her creation to the woman of the house, assuring her that it was easy to be a great seamstress with the help of the amazing Singer sewing machine. "I made this for you last night," she remarked nonchalantly with a smile. "It was no trouble at all." Mom sold quite a few sewing machines that way.

And she taught me that knocking on doors when you had something good to present was an effective means of marketing. That lesson would come in handy up the road years later.

CHAPTER 5

Depression, Dust, and Dawson's

I was six years old when the stock market crashed, ushering in the Great Depression that devastated an entire generation of Americans. Adding insult to injury, in Kansas a severe drought made the parched land even more vulnerable to the wicked winds that howled across the region. For most of my youth, it seemed rainfall could be measured by the thimblefuls. Russell sat right in the center of the "dust bowl," and the topsoil of many of the farms literally turned to powder and blew away, taking with it the hopes and dreams of some of Kansas's most hardworking families.

If you've never experienced a dust storm, consider yourself lucky. It's hard to imagine how horrible and frightening one can be. I remember as kids, we could look out the windows at school and see the large, ominous clouds taking shape in the sky and shrouding the sunlight. It was a weird feeling, knowing that according to the clock, it was the middle of the day, but because of the accumulating dust, the sky had already darkened, turning a dreary, dirty, brackish brown.

Parents began arriving at the school to retrieve their children while they could still see well enough to find the building. Once it became obvious that the storm was heading our way, schools let out early. My sisters and Kenny and I ran home to help Mom and Dad seal up the house as best we could. We'd fill the bathtub with water, soak some towels, and then pack them around the windowsills and doorways, hoping to keep out some of the dust. The wet towels helped, but nothing sealed the dust out completely. Before long, a thin layer of silt covered everything in the house, including any food or water left unprotected.

I delivered the daily **Salina Journal** on my bicycle each day, and occasionally a dust storm blew up so fast I hardly had time to get back home. Blind to the road before me, I'd wet a handkerchief and wrap it around my mouth and nose to keep from inhaling the dirt.

Although I was still too young to be on the team, I heard plenty of stories about a Russell high school basketball game played at nearby Hays High during a dust storm. Apparently, the floor had to be swept every five or ten minutes, just so the players could see the lines on the court.

It was not uncommon during the worst of the dust season for cars to be covered, and like huge, dirty snowdrifts, sometimes even low-rise buildings were obliterated from sight.

Shrieking like a monster in a B horror movie, the dust-saturated winds soon turned into killers. Without rain, there were no crops; without crops, the farms were worthless. Land prices dropped like rocks. People who were struggling to make ends meet even before the onslaught of the dust bowl period known as the Dirty Thirties now battled for sheer survival. The dust killed everything it covered. The crops died, trees, flowers, and other vegetation died, and people died—some from dust pneumonia, some from despair and self-inflicted gunshot wounds. Everyone in town knew of a nearby farmer who had gone to the bank and been turned down for a loan. Everybody in Russell had been broke at one time or another.

By the mid 1930s things got so bad that Mom and Dad decided to move into the concrete-floored basement of our house and rent out the upstairs to some oil people, simply to make ends meet. Dad installed a bathroom and rigged up a shower in which we could bathe and a stove so Mom could cook. We lived in the basement for several years. To help out, everyone in the family worked. I delivered newspapers, mowed lawns when there was grass to be cut, raked leaves, shoveled snow—anything to earn a few extra nickels. My sisters and my brother did the same.

Throughout the Depression years, talk abounded around Russell about President Franklin Delano Roosevelt's New Deal, the programs by which the government hoped to pull the country out of the economic trough and get people back to work. Occasionally we gathered around the Philco to listen to Roosevelt's optimistic predictions of an economic turnaround. No doubt the New Deal was a topic of conversation with Dad and his friends over at the grain elevator, but we never really talked much about politics at home. Dad much preferred talking sports to politics.

Neither of my parents was politically active, although they were registered Republicans, as were most Russell County residents. But when Cliff Holland, an American Legion buddy of Dad's, ran for Congress as a Democrat, Mom and Dad had no qualms about switching parties. Whether out of appreciation for Roosevelt or loyalty to their friend, they remained registered Democrats until the day I entered politics.

Similarly, although my parents' religious faith permeated their lifestyles and personalities, they were not particularly active in our local congregation. They attended church services whenever they could, but more often than not Mom and Dad worked on Sunday mornings. This didn't prevent their sending us kids to Trinity United Methodist Church in Russell. I admired the pastor, Reverend William H. Jenkins, a calm, soft-spoken, good man who practiced what he preached. In Sunday school classes, I learned the importance of faith in God, but also about truth, honesty, right and wrong, loyalty to friends, doing one's duty, trusting and believing in good in the face of overwhelming odds. Even more important, Mom and Dad taught me the ways of the straight and narrow, not merely with words, but in the way they lived.

I guess I've always been known as a rather straight-and-narrow sort of guy, even though I've not been outspoken about my faith. And I'd be quick to admit that I didn't always remain on that straight-and-narrow path—much to my mother's dismay.

Nevertheless, having a foundation of faith in God, believing that life matters, that there's a bigger plan in play than simply that which we can see with our human eyes—such truths were instilled in me from an early age. Having the faith to endure would one day become one of the most significant factors in my life.

Even as a boy, I rarely walked anywhere; I preferred to run. Sometimes, early in the morning before the sun crept up in the sky, I'd trot over to my friend Dean Krug's house, on the outskirts of town. Dean's dad, George, was a carpenter by trade, but the family kept a few milk cows, too. At dawn, Dean and I'd do the milking, and then whatever milk Dean's family didn't need we'd sell to the grocery store for a nickel a quart. A two-and-a-half-cent split sounds silly today, but in the 1930s that was **real** money for a kid in Russell.

My brother, Kenny, often joined Dean and me, and the three of us worked all sorts of odd jobs to make a few cents. Usually, I was the one who got the job, and then I persuaded Kenny and Dean to help. We pulled weeds, dug out dandelions, washed cars, and delivered grocery store handbills announcing all the weekly specials—one of the most efficient ways of advertising back then. Passing out the handbills was also one of the most lucrative jobs for us. There were about eight hundred homes in Russell by then, and Mr. Holzer, the grocery store manager, expected us to place a handbill in every one of them. The job paid a whopping two dollars. It took hours for the three of us to deliver all the ads, but even after splitting the proceeds, we each walked away with more cash than some entire families had at the time.

As the economy gradually improved, we ventured into more entrepreneurial waters and tried our hands at selling the infamous cure-all Cloverine Salve. I was never quite certain what the stuff in those green-and-white tins really remedied, but according to the official spiel, that salve could improve just about anything. All I knew was that we bought it twelve tins to a sleeve, and if we sold all twelve—which sometimes took several weeks to a month—at twenty-five cents a tin, we could make fifty cents per shipment. Of course, the most likely prospects were neighbors and relatives, and by the time I turned twelve, our neighbors and relatives probably owned enough Cloverine Salve to grease a truck.

No doubt, most of the people who knew me were thrilled when I turned thirteen and got a job down at Dawson's Drugstore. At least they wouldn't have to buy any more salve.

Located just a stone's throw away from where Dad worked on Main Street, Dawson's was the gathering place of Russell. Sooner or later, everyone in Russell came through the doors of the old-fashioned drugstore. Some came to have prescriptions filled, others came for coffee and conversation, and many came for the candy, sodas, and ice cream.

The business was owned by old "Dutch" Dawson, who still showed up for work every day, but spent most of his time keeping an eye on things from a back booth. Dutch's oldest son, Ernie, was the pharmacist. A quiet sort of

fellow, Ernie kept busy in the back filling prescriptions. The real flavor of the place came
from Dutch's younger sons, Chet and Bub. The
boys believed in having fun at work, and they
possessed two of Russell's quickest wits and
tartest tongues, which they tempered slightly,
smoothing the rough edges off their sarcasm
enough to make it palatable for most of the
townspeople who frequented their place. The
Dawson brothers poked one-liners at everyone
who came through. Most people took their
quips in the good-natured spirit with which they
were intended.

The Dawsons peppered their employees with
jabs as well. Standing next to a customer, Chet
would call across the room to a new employee,
"Hey, kid. Come over here. This man was clean-
shaven when he came in here, and now he's got
an inch growth of beard while waiting for you to
take his order!"

All day long, you could find older men holding court at Dawson's, sitting at the wooden tables under the ceiling fans, engaged in lively
banter, arguing about this, complaining about
that, but mostly just whiling away the hours,
working crossword puzzles or playing checkers.
At night, after the movie show let out at the

Dream Theater, across the street, or when the nearby Mecca, our other theater in Russell, emptied out, moviegoers ambled on over to Dawson's to top off their evening with a bit of sweetness. Soon the kids poured in after a hot night at the roller-skating rink, and the stools and standing area at the soda fountain filled to capacity. Dawson's was the place to meet.

The Dawsons hired me as a soda jerk, which was right up my alley. I worked most week nights and almost every Saturday morning throughout my high school years. For a dollar a day, I dished up ice cream, whipped up milkshakes, and served chocolate malts and Green Rivers, a local favorite drink made with lime. Dawson's made the best ice cream sodas anywhere around. Bub liked to boast that he had secret recipes, including chocolate imported from England. Maybe so; it could just as well have come from down the street. I didn't care. I was too busy indulging my sweet tooth.

Sometimes when the kids from school came in, I'd flip their ice cream in the air, landing it right in the cup or glass, before adding the soda, cherry, and the other toppings. The Dawsons didn't mind my fooling around. To them, it added to the show—as long as I didn't drop any of their precious product. I'd take a kid's order and bark out, "One Green River, coming up!" Turning to the customer, I'd drop my chin and raise my eyebrows, as I asked seriously, just as Bub and Chet had taught me, "Do you want the flip in it?"

"Oh, yeah," the kid would say.

I'd toss a scoop of ice cream toward the ceiling, sending it into slow-motion somersaults, watching it flip over and over, timing it just right, letting it fall as long as I dared before swooping it into the cup, under the Green River soda spigot, and onto the counter in front of the customer, seemingly all in one fluid motion. Chet or Bub would stand off to one side of the drugstore, beaming.

I got along well with the Dawsons, and quickly picked up their patter and droll sense of humor. Before long I was dishing up friendly wisecracks as well as ice cream sodas. At ten o'clock each night, Dutch closed up and checked the registers, before heading home. The rest of us stayed behind to clean the entire place, getting everything ready to open the next morning. We got to be good friends, especially Chet, Bub, and I, who were closer in age. When everything was spotless, I doffed my white apron and ran from Main Street to our home, on Maple Street, where Mom always had something ready for me to eat, topped off with a special cake or pie made by Gloria or Norma Jean.

Good thing I enjoyed running. Otherwise, between the ice cream samples at Dawson's and Mom's late-night dinners, I'd have weighed three hundred pounds. Occupational hazard, I guess. Still, I managed to keep myself in pretty good shape.

I got up early every morning to go running, pounding Russell's sidewalks long before jogging was fashionable. Part of it was sheer enjoyment; the other part involved high school athletics. I played football and basketball for Russell High, and ran the 440 and 880 middle-distance runs on our track team, even setting a record in the half mile (although the record was broken shortly thereafter). Not really a naturally gifted athlete, I was nonetheless a competitive person; I wanted to win, and I would do all that was necessary to be in the best physical shape possible.

To help build our strength, my brother Kenny and I made our own set of barbells out of cement attached to either end of an iron bar. We lifted those weights, grimacing all the while, every chance we could. I wanted to be ready when oppportunity and I crossed paths.

I was a three-sport letterman at Russell. When the hot August winds started blowing, I put on the pads and went out to football prac- tice. I started on offense as an end, although Coaches Dean Skaer and George Baxter weren't much interested in throwing the ball. They pre- ferred grinding out the yardage the old-fash- ioned way, on the ground. In our senior year, the Russell High Broncos went undefeated, racking up nine straight wins on the gridiron. Much of our success was due to Bud Smith, son of an oil family who had moved to Russell in 1937 to run the supply store, selling the pipe and pumping units used in the oil wells. With movie-star good looks and dazzling athletic skills, Bud could an- ticipate a great future. He and I became close friends and a strong one-two combination on the football field. He was the quarterback; I was the end. He made my job easy—he threw the touchdown passes, and I just had to catch them.

We were joined by Phil Ruppenthal, whose father was a Democratic judge. Phil spent most of his free time in the public library reading everything he could get his hands on. He was the intellectual of our group, and later went to

Harvard on an academic scholarship. Phil survived World War II, but sadly, he died much too young, of a heart attack.

Another of our good friends was Roland Rautenstraus, son of a local Lutheran minister. Wherever anyone saw Bud, Phil, or me, Roland could be found as well.

A tackle on the football team, Adolph Reisig, was another good friend from high school who would later play a key role in my life. We ate, drank, and slept football from August through November.

But basketball was my first love. By my senior year, I was pushing 192 pounds, and was nearly six feet, two inches tall. Meanwhile, the game itself had changed in recent years. Invented by James Naismith, basketball had originally required a jump ball after every basket. But in 1937, the rules were altered so when one team scored, the other team automatically got to take the ball out, and start down court toward its goal. The result was much more running during the game—which was perfect for me. I didn't score a lot of points, but I relished any chance to help my teammates score. I must have done it fairly well, too, as I was the only guy from Russell High selected for the Union Pacific All-

Stars, a squad chosen by conference coaches from schools up and down the railroad line that ran through Kansas. (A lot of credit goes to my coach, Harold Elliot.)

My athletic career, such as it was, greatly enhanced my popularity at Dawson's, much to the delight of the Dawson brothers and their dad. They preferred to hire the popular kids in school, especially the athletes, because that further labeled the drugstore as the hot spot in town—or the cool spot, depending on how old you were. Regardless of age, sports talk was always on tap at Dawsons, and especially when Kansas State University, located in Manhattan, played the University of Kansas, located in Lawrence. The Dawsons were big K-State fans, decking out the drugstore in KSU's purple-and-white pennants and other sports paraphernalia. Eventually they permitted a few red-and-blue University of Kansas items to be placed on the other side of the drugstore, if for no other reason than to egg on the KU fans who came in to join in the heckling.

I was a reasonably good student in those days—pulling much better grades than I would in college—enough to gain admission to the National Honor Society. I was sports editor of our school newspaper, **The Pony Express**, and president of Hi-Y, an organization of young Christians. For all this, I never regarded popularity as something to be sought; it was merely a by-product of my long-standing dream of becoming a great athlete. But apparently the girls liked me, because I was tall, had dark, wavy hair, and spoke politely to them. Near the close of my senior year, the members of the Russell High Girls Reserve voted me as their "Ideal Boy."

I once even "modeled" some clothes for a department store spring fashion show. Despite my natural shyness, I seemed to connect with the audience. I never hitched up with an "Ideal Girl," however. Not that I was uninterested. I was simply too busy. After my studies, sports, and doing my job at Dawson's till nearly eleven o'clock each night, my social life was low on the priority list.

Besides, I'd have plenty of time to meet girls once I got to college. **College?** Nobody in my family had ever gone to college. Yet more and more, I noticed that the most successful and respected customers—those who really had clout in Dawson's—were doctors. Moreover, medical men such as Dr. Koerber, our family physician, and Dr. White, another local practitioner I admired, contributed enormously to the overall life of the community.

I developed a plan: somehow I was going to earn enough money to attend college. I'd go on to med school, eventually become one of those doctors who didn't have to worry about which way the wind blew or how much rain we were likely to get this summer. **I'd have some sense of security, and the satisfaction of doing something of significance to help other people, while providing for myself—and my wife and kids. . . . Ha! Listen to me, thinking about a wife and kids. . . . I've hardly been out on a real date.**

As my ambition took root, so did my conviction that college held the key to future success. Even my graduation ceremonies seemed to point in that direction. In June 1941, I sat with the graduating seniors of Russell High and listened intently as Reverend Rautenstraus reminded us of the biblical injunction, "Narrow is the gate and strait the way that leads to life." For me, the narrow gate led straight toward the University of Kansas.

CHAPTER 6

College Man

I thought for a moment that Bub Dawson was going to choke when I broke the news to him. "Why in the world would you want to go to KU?" he asked incredulously. "Snob Hill, of all places!"

Built on the lush, green-grass slopes of Mount Oread, Kansas University did indeed look down on the rest of the state, especially the corn and wheat fields of eastern Kansas. Conversely, people who had no intention of ever attending KU often referred to the school as Snob Hill.

"I know, Bub. I know. But it's my best chance. You know I got that letter from Phog Allen. He seems to think I can make the basketball team."

Phog Allen was KU's legendary basketball coach, and possibly the best-loved figure in the state. For Kansans, college basketball has always been king, so much so that KU had brought in the game's inventor, James Naismith, as the university's athletic director. Naismith mentored some of the greatest coaches and players ever to step on the hardwood. Adolph Rupp, idol of the University of Kentucky basketball program— the man for whom the Kentucky Wildcats' arena was named, and whose legacy of victory casts a long shadow even now—played basketball for James Naismith at the University of Kansas. So did Forrest C. "Phog" Allen, the man that many Kansans regarded as the father of modern basketball.

I had met Allen's son, Milton—we called him "Mitt"—when I worked for the Kaw Pipeline that summer. He was playing for an amateur basketball team in Russell. When his dad came to visit, Mitt told him about his friend Bob Dole who was a hardworking, ambitious guy who worked over at Dawson's Drugstore—and, oh, he's not a bad basketball player, either.

I about died on the spot when the most respected coach in college basketball walked in to Dawson's with his son. Phog Allen stepped up to the soda jerk counter, his son introduced us, and Phog shook my hand.

"I hear you're pretty good with a basketball," Phog boomed, his voice indeed resonating like a foghorn on a ship. Every eye in Dawson's turned in my direction.

"Er, ahh, yessir," I mustered.

"Why don't you come on over to KU next fall? We have some spots on the freshman team; you might fit in well."

I thought for a moment of my friends Chet and Bub, and how they despised KU. **But come on, guys, this is Phog Allen.** "Ah . . . aghh . . . yessir. That would be fine."

A few weeks later, I received a letter from Phog Allen reiterating the message he had conveyed at Dawson's. "Great to meet ya. Look forward to seeing you in Lawrence in the fall." The letter contained no scholarship offers, but it was a bona fide invitation from Phog Allen to try out for his team. Only a fool would have passed up such an opportunity.

Or someone with no money.

I talked with Harold Dumler, another Russell friend, who was a few years older and ready to begin his senior year at KU. His fiancée had been my Spanish teacher at Russell High. Harold strongly encouraged me to consider enrolling at the University of Kansas, where he was a member of the Kappa Sigma fraternity. He suggested that if I joined the frat, I might be able to work at the house, and earn enough money to help keep myself in school. Harold spoke to his fraternity brothers about me in glowing terms, telling them what a great athlete I was, as well as a good student, popular, and an all-round good guy with a great future—just the kind of guy they were looking for to be a Kappa Sig.

I also broached the idea of going to college to Mom and Dad. They were surprisingly supportive, considering that neither they nor any of their relatives had ever gone to college.

But I also knew how tight things were financially at home, and how hard Mom and Dad were working already to support the family. I didn't want my education to be a burden on them. Yet I knew it would be. Not only would going to KU cost me money, but it would erase what little extra financial assistance I was contributing to our family budget.

Nevertheless, Mom and Dad encouraged me to do what I felt I had to do to make my mark in the world. "Don't worry about the money," Mom said. "We'll find the money if it comes to that."

I started scraping up my pennies, but I still didn't have enough to pay the tuition at KU. I sought help from George Deines, a local banker. Mr. Deines was willing to loan me the three hundred dollars I needed to get started. Along with the check came some free advice. "Wear a hat," Mr. Deines told me. "Gotta get a hat, if you're going to be a college man." The banker wasn't concerned about my catching a head cold; to him, a hat signified maturity, stability, someone you could trust to repay a loan of three hundred dollars. Mr. Deines wore a hat, so he thought that anyone who amounted to anything should also wear a hat. He was a fine man who helped many young citizens of Russell, and I will always be grateful to him.

I went out and bought a full-brimmed felt hat. I don't think I wore it more than a few times, mostly when I walked by the front windows of the bank. I wanted to make sure that Mr. Deines knew that I'd gotten a hat, so he could feel better about his investment. After all, I had years of college ahead of me; even though I planned to work my way through school, someday I might need Mr. Deines's help again.

Everyone was so friendly at college that I couldn't for the life of me figure out why people referred to the University of Kansas as Snob Hill. Wrapped around the rolling hills of Lawrence, a midsize town about forty miles west of Kansas City, the KU campus seemed to sprawl in every direction. **I'm going to love running here,** I thought, **not just on the flat stretches, but up and down the hillsides.**

Soon four thousand students would swarm the campus—nearly the equivalent of the entire population of Russell—but because I was trying out for the KU football team, I had traveled to Lawrence in late August 1941, a few weeks before classes were scheduled to begin. True to his word, Harold Dumler, my hometown friend, helped ease the transition from life in Russell to life in Lawrence. Harold introduced me to the guys in the Kappa Sigma fraternity, and although several other frats "rushed" me, my loyalties were to Harold. He helped me get a job waiting on tables in the Kappa Sig fraternity house dining room. My pay was $12.50 per month and all that I could eat. I thought that was a pretty good deal. Better yet, as a member of the fraternity, I could live less expensively at the frat house.

Bud Smith went with me to check out the Kappa Sigma house. "Go ahead," Harold told us as he showed us the room he shared with another frat brother. "Take a look around. I'll be back in a few minutes."

Bud and I couldn't resist. All those beds, with time on our hands. We started with Harold's bed, stripping it down and remaking it—"short-sheeting" it—so when Harold climbed into bed that night, he'd have about enough room to crawl in up to his knees. Then to make sure we made a good impression on the rest of the guys, we went around and short-sheeted several other frat brothers' beds. Talk about how to make friends and influence people!

I guess the guys admired my brashness, because they didn't rescind Bud's and my invitations to join. Either that or they figured that they'd get even during "hell week," the final test on the bumpy road toward membership. Like most fraternities, the Kappa Sigs had a pledge program in which the new guys—the pledges—had to prove they were Kappa Sig material. Usually this involved doing silly things like wearing burlap sacks for underwear or working at demeaning chores and assignments, such as scrubbing the restroom floors with a tooth-brush. If a pledge didn't do things just so, he was given a few whacks with the fraternity paddle, a flat board made from a barrel slat. It was all in good fun, and nothing like some of the dangerous hazing kids have experienced on campuses in more recent years.

One day I was in the house with a couple of the active brothers, and I said, "I've heard so much about those paddles, I better find out how bad it's gonna be." I turned to one of the older frat brothers and said, "Come on. Give me your best shot."

The brother thought I was loony at first, but he eventually complied with my request, to the great amusement of several guys looking on. I bent over, and the brother hauled back and walloped me with the paddle as hard as he could swing it. He lifted me right off my feet and sent me lurching forward. My rear end stung like it was on fire. But I wasn't going to give him the pleasure of knowing it hurt.

I turned around, nodded at him, and said, "Well, that wasn't so bad."

Very few hit me with the paddle after that.

I figured turnabout was fair play, though. One of the brothers had a big Harley Davidson motorcycle parked outside the house. I rounded up a couple of the guys, and we hauled that cycle all the way up to a third-floor bedroom.

When the brother got back to the house, there was his motorcycle, upstairs waiting for him. We made ourselves scarce and refused to help him get the thing down.

Most of fraternity house life was rather tame, however, especially compared with the images of fraternities later popularized by movies in the 1970s. Granted, we did enjoy our share of parties. But unlike the rowdy, raucous behavior of frat brothers in the movies, our fraternity was more like a boys' club on campus. We wore sports jackets and ties to functions. We received lessons in matters such as dinner-table etiquette, conducted by our matronly housemother. And first-year students in our frat had to follow strict study hours, despite our fooling around. That was okay with me. I could tell early on that I was going to need all the study time I could get.

Right before classes started, in addition to my waiter's job, I picked up a second job to supplement my modest income. Bud Smith and I delivered milk in Lawrence on Saturday and Sunday mornings. The idea of getting up on the weekend before the birds didn't thrill me, but I needed the extra spending money.

I was excited about beginning my studies,

and a little nervous, too. I planned to enroll in a premed program, but most of my initial classes would be general education courses. The night before classes began, I wrote home to my family:

Dear Folks,
   School starts tomorrow and I guess unless something happens, I'll be taking 5 hours of German, 5 hours of Economics, and 2 hours of Rhetoric.
   I started waiting tables yesterday; it's kind of hard at first, but I guess it will become easier as we get onto it. I guess Bud and I have to start our milk job tomorrow, which I don't like. We never get to bed before 12:00 and getting up at 5:00 is not very easy.
   The freshman football team scrimmaged the varsity yesterday for about an hour. I was playing Right End on the freshman first team, and I mean those varsity boys are really tough. I spent half my time picking myself up. I have a black eye, too, as a result of the scrimmage. I think I have a pretty good chance of making the team permanently.

> **When are you going to send my clothes up? I haven't had a clean shirt for two days. I've been waiting for them all week.**
>
> **Can't anybody write in our family? I haven't gotten a letter from home yet. Well, it's 9:10 AM and I have to enroll at 11:00 so I'd better stop.**
>
> **Love,**
> **Bob**
> **Send those clothes tonight!**

It was the first of many letters I'd write in a rush to my family members. I'd never been so far away from home for such a long time, so I wanted to keep in touch with everyone. Plus, I knew Mom especially would be worried about me, and Dad would be concerned that I was fooling around too much, and Kenny, Gloria, and Norma Jean would be interested in what college life was like.

Looking back at some of our letters from that period, I'm struck by how closely knit our family unit really was. We were always a warm family, though not effusive in expressing our love and concern for one another. Nor were we accustomed to outward expressions of affection such as public hugging and kissing. Like many men of my generation, I wasn't too good at expressing love out loud. But when I went away to school, suddenly the distance between us stimulated a stronger desire to communicate, at least on paper. Many of our letters during that time dealt with mundane matters such as my bills, laundry, grades, and the various high school and college sports scores. Mostly they were about keeping in touch.

We really missed one another, although rarely did any of us put those sentiments on paper. Mom and Dad were never much for flowery words and compliments, but we kids knew they genuinely loved each other and loved us, too. Similarly, Dad didn't spend a lot of time patting us on the back for doing the expected. He didn't say things such as "Good job, son." When we did our work with excellence, Dad might acknowledge it with a "Pretty good," and that meant more to me than phrases such as "Great job!" "Incredible!" "Fabulous!" and other accolades that tend to lose their significance with a lot of vain repetitions.

I also noticed that in my letters, I was always genuinely respectful of my parents. I'd tease and cajole my sisters and brother, and poke fun occasionally at my parents, but I never crossed the line of being disrespectful, expressing disgruntlement to them, or casting any aspersions on them for not being better able to help me out financially as I struggled to make ends meet. I knew they were struggling, too. Nevertheless, sometimes it's all too easy to say nasty things to the people who love you the most, simply because you know that they will still love you the next day, despite how badly you've hurt them with your words. But thankfully, that didn't happen between my family and me.

At KU, I continued my early-morning running regimen before the other guys in the frat house awakened. In addition to my workouts with the football team, I also continued working out with my homemade barbells, which I'd persuaded Harold Dumler to transport from Russell to Lawrence in his car, along with some of my other belongings. I kept the weights in the frat house and used them regularly, working to tone my body and build my strength.

The one thing I had trouble managing during my first semester at KU was my laundry. Laundromats didn't exist at that time, so all the Kappa Sigs had to take turns using the machine at the frat house. Rather than doing the wash myself, I either paid to have my clothes laundered or I sent them home with friends going back and forth between Lawrence and Russell. I explained my dilemma to my mom in a letter written in September 1941. This letter is typical of many that I wrote to her during my first weeks of higher education:

Dear Mom,
  I hated to send all this laundry home, but it costs too much down here; shirts are $.15, shorts and short-shirts are .06; towels, .05, and wash rags .02, socks .03, and that adds up pretty fast when you have as little money as I have. . . . I have to pay $50.00 for fees this afternoon, and I've spent about $12.00 for books and I still need a German book that will cost $2.50. Altogether I've spent about $70, which leaves me only about $35.00 left in the bank. Cleaning is high here, too. Pants are .50 and suits are .65. . . .
  Sometime within the next month or two, when you're not too busy, send me some cookies or something.

> **I'm also sending a list of the freshman rules, so that you won't have to worry about me.**
>
> **Love,**
> **Bob**

One mention of cookies, of course, was all it took for my mother. The next week I received a box containing some of my clothes, and another stuffed full of her home-baked oatmeal-raisin cookies.

It was now five weeks into the school year, and I'd found a new job working from one-thirty until three-thirty each afternoon, for forty-five cents per hour, which I considered plenty good. I could have worked longer and earned more, but then I'd have had to quit the football team and I didn't want to do that. I was doing well on the freshman squad's first team, and although it was rough scrimmaging the varsity every night, I loved it.

Getting acclimated to the university routine, I still relished news from home. In a letter addressed to the entire family, I gently castigated my brother for not writing to me. "Did Kenny break his arm in football practice?" I asked. "He must have done something. He hasn't written a word in five weeks. Tell him to keep trying and I'll show him how football is really played in a couple of weeks when I come down to visit and catch up on my sleep. I studied till four A.M. Wednesday. Well, it's 1:15 and time I started for work. I'm working in a greenhouse. Have Kenny write when he gets his arm out of the cast. Ha-ha! Dad, you might write yourself when your back gets limbered up."

It was a theme I often repeated: "Kenny, write me and tell me how you're doing in football and in school. What dame are you going with?"

"Norma Jean, write and tell me about the tooth they pulled, and also how you're doing in school. . . . How much do you weigh now—110 or 150?"

My younger sister failed to see the humor in my teasing.

"Sister [Gloria], how are you and Harold getting along?" Harold was the fellow she was dating at the time. "I've had five dates. One was with a twin . . . and she's plenty cute, I think. Don't forget to write me soon. Love, Bob."

In October, I received a real surprise. The KU football team was playing Nebraska, and our coach decided to take the freshmen along. "We go Saturday morning at 6:00 A.M. and come back the same day," I informed the folks. "We're going in a special train. They pay for everything, our tickets, our food, and our train fare. We should have a lot of fun; KU will probably get beat though."

In that same letter, I couldn't pass up an opportunity to tease my buddies back at Dawson's Drugstore. "I got a letter from Chet the other day and he said that the Dole family won the World Series pool, as usual. The next time he razzes you, ask him how Kansas State did in their game with Northwestern last Saturday!"

The trip to Nebraska was great fun all right. Unfortunately, our team didn't play so well against the tough Cornhuskers. The following week, when writing to the family, I admitted: "About all I can say about the Nebraska-Kansas game is that KU had eleven men on the field . . . even though it didn't look like it."

During my first few months at school, I was constantly on the lookout for some additional means of earning money to support myself. Passing through the second floor of the frat house one day, I noticed some vacant space. I had a brainstorm. Back home, following the oil boom, many businesses had installed five-cent and ten-cent slot machines. Technically, the machines were probably illegal, but it seemed every store in Russell had one tucked in a corner somewhere. My dad even had a slot machine at the egg-and-cream station when I was just a little boy. I wouldn't have been surprised to learn that the county sheriff's office had one, too.

I wrote to Dad asking him what he thought about the idea of our installing a slot machine in the fraternity house. He responded positively, and promised to look into the possibilities. When I broached the subject to some of the frat officers, they thought it was a wonderful way for the fraternity to make a little extra money, while doling out some small prize payouts to winners. Of course, as the entrepreneur who serviced the slot machine, I planned to make a few pennies, as well.

"I asked our treasurer about putting a slot machine on the second floor, and he thought it would be a swell idea," I gushed to my folks in a letter home, just before Thanksgiving. "He believes I could make around $5 a week. If I could do this, I could quit one of my jobs and have more time for study."

I needed the time as much as or more than the money. I was doing well in rhetoric, and had finally pulled my economics grade up near to a

B average, but I was still struggling with German, barely eking out a C. I never did get around to installing a slot machine in the frat house—it just didn't happen.

By the end of November, I had scraped together enough money to buy a secondhand divan for seven dollars. The sofa made my room look flashy, and I was proud of it. I wanted everything to be perfect for when I was officially initiated into the Kappa Sigma fraternity—just as soon as I got my grades up. I wrote to my family regarding my preparations for the big event:

> **Roy, one of my fraternity brothers, gave me a suit yesterday. I sent it down to have it altered; the coat fits fine, but the pants are too big in the waist and seat. He claims he paid around $60 for it, but you can't tell much about what he says. He told me to take it, and if I ever have any extra money, I can give it to him. All in all, it's a pretty good deal.**

One other "luxury" item found its way into my room at the frat house during my early days at KU. It was a used phonograph machine that played 78-rpm heavy vinyl disks, with one song on each side. Soon I was listening to Tommy Dorsey and his band, to Glenn Miller performing "In the Mood," as well as to a new artist who seemed to have quite a lot of potential. His name was Frank Sinatra. Lacking funds to buy a lot of records, I relied on my fraternity brothers who brought some records in to share. Mostly I just played the ones I had over and over. I could never have imagined what a key role that phonograph machine would play in my life in the years to come.

I was in the fraternity house on December 7, 1941, when the quiet, peaceful Sunday afternoon was shattered by a news flash on the radio:

> We interrupt this broadcast to bring you this important bulletin from the United Press. Flash. Washington. White House announces Japanese attack on Pearl Harbor.

I could hardly believe my ears. We'd been hearing reports in the news about the war in Europe, how Hitler had overrun Poland in 1939 and had conned Neville Chamberlain into appeasing him, all the while making plans to rampage through Europe, subduing millions of people under his vicious regime. But that war seemed so far away. I hate to admit it, but most of us guys in the Kappa Sigma house in 1941 were more concerned about who won the big game or getting a date for the Theta's sorority party the next weekend than what was happening on the war front.

But our indifference, and our ignorance, vanished in an afternoon. Now our country had been attacked. By Japan of all people. Who did they think they were? Who did they think **we** were, that they could so boldly bomb our military bases in Hawaii?

The following day, in what would be known as one of his most famous speeches, the somber voice of President Franklin Delano Roosevelt informed the joint session of Congress as well as the rest of the country:

Yesterday, December 7, 1941—a date which will live in infamy—the United States of America was suddenly and deliberately attacked by naval and air forces of the Empire of Japan. The United States was at peace with that Nation and, at the solicitation of Japan, was still in conversation with its Government and its Emperor looking toward the maintenance of peace in the Pacific. . . . It will be recorded that the distance of Hawaii from Japan makes it obvious that the attack was deliberately planned many days or even weeks ago. During the intervening time the Japanese Government has deliberately sought to deceive the United States by false statements and expressions of hope for continued peace.

President Roosevelt went on to say, "The people of the United States have already formed their opinions and well understand the implications to the very life and safety of our Nation. As Commander in Chief of the Army and Navy I have directed that all measures be taken for our defense. But always will our whole Nation remember the character of the onslaught against us."

I certainly didn't want to go to war—nobody in his right mind would—but as my frat brothers and I crowded around radios listening to the president's words, a sense of patriotism we'd never fully experienced before began to well in all of us. We could sense the resolve in Roosevelt's voice as he concluded, "No matter how long it may take us to overcome this premeditated invasion, the American people in their righteous might will win through to absolute victory. I believe that I interpret the will of the Congress and of the people when I assert that we will not only defend ourselves to the uttermost but will make it very certain that this form of treachery shall never again endanger us.

"Hostilities exist. There is no blinking at the fact that our people, our territory, and our interests are in grave danger. With confidence in our armed forces—with the unbounding determination of our people—we will gain the inevitable triumph—so help us God."

The president then asked the Congress for a declaration of war with Japan. Within twenty-four hours, the United States was officially in the Second World War.

On campus, we tried to carry on as normal, concentrating as best we could on our studies and class work, planning fraternity parties and other events. But it was all a bit surreal. For example, I had gone out for the basketball team—and Phog Allen had picked me to be a part of the freshman team—yet somehow this didn't seem important now. I prepared for semester final exams, but it all felt hollow. Some of my friends had already enlisted in the military, and many others were on their way. I think we all knew instinctively that our college days would soon be over.

As we listened to the reports about Pearl Harbor, most of us considered enlisting immediately, and some did, simply dropping out of college and seeking out the nearest courthouse or place of enlistment they could find. I considered it myself. A day or so after the president's speech, I wrote home:

Dear Folks,
  I suppose you, like everyone else, are huddled around the radio listening to war developments in the battle with Japan. All day yesterday, boys in the house sat around their rooms, waiting for news bulletins to come in. There are three or four boys who are considering enlisting this week in the Army Air Corps. It might be a good thing for me to do; at least it would be better than spending your money at school!

Around the fraternity house, emotions ran the gamut from shock to anger to futility. Of course, some of the guys engaged in nervous bombast, boasting how they were going to go over there and kick some "Jap butts" or beat those "Krauts" into submission. Everyone was furious that our country had been so flagrantly attacked, although at that time, we didn't know the half of it. Reports of the damage and death toll at Pearl Harbor were sketchy at best. Some guys had an "Aw, what's the use?" attitude about going to class or continuing their academic pursuits. They were resigned to the fact that our nation was now at war, and it was only a matter of time until they'd be in uniform, and probably on the front lines.

For a while, I decided to maintain the status quo. I hoped to be able to finish out the semester, and possibly even my full freshman year, before being called into service. There was no real question about whether I'd be going; the only questions were when, where, and in what branch of the military I would serve.

I came home for Christmas, and our family relished the time together. Of course, Gloria, Kenny, and Norma Jean wanted to hear all about the university, which pretty girls I had dated, and what good-looking fellows I might send my sisters' way. Kenny and I talked sports, both Russell High and KU, but the war news cast a pall over any other sounds coming from the old Philco radio, or anything that we'd read in the **Salina Journal**. Mom and Dad freely discussed with Kenny and me the real possibility that we'd be called into service soon. Dad recounted stories of his service time spent in Texas, when he'd been itching to get overseas. We laughed a little nervously. Although I'd gone hunting with Dad, I wasn't a great marksman. The idea of toting a gun in Japan or Germany was a bit daunting. Mom hoped that if Kenny and I were summoned, we might serve as Dad did, on the home front.

For all its harsh realities, the war still had a newsreel feel to it. It was something that was happening far away, a film clip that we watched before the feature came on at the Dream Theater. That picture soon faded given the number of Russell boys who had already enlisted in the army—including Bub Dawson, Eugene Ruff, a neighbor who lived across the street from us and was a couple years older than I, and some of my other hometown friends. The war touched us all, one person at a time.

I returned to Lawrence at the beginning of 1942 to finish out the semester and to take my final exams near the end of the month. It was tough to study with the constant flow of war news in the background. I resumed my habit of writing home every few days, my letters reflecting the conundrum with which we lived daily— one that balanced the mundane matters of everyday life with the ominous specter of global conflict.

Hello everybody, how's things in Russell? I just finished reading the *Russell Record*. I see that Russell really trampled Quinter. I also saw that Mrs. Doran Dole entertained her bridge club . . .

We've lost three boys since Christmas. Two joined the Army and one quit school. We'll probably lose around five or ten more when the first semester is finished . . . I guess I'll plan on going on to school next semester unless they change the requirements for the Army Air Corps. Bud was told today that anyone who is eighteen with a high school education

and physically fit could become an air pilot. Well, I guess I better start studying if I want to pass that Economics final exam.

I joined the track team that spring, won some races, and contributed to the success of our team; I just missed setting a record in the indoor 440-yard dash. Football, basketball, and track, that's what I lived for—that, and a good chocolate milkshake.

Looking back, it's amazing how naïve I was, thinking, hoping, praying that the war would come to a speedy conclusion, that it wouldn't disrupt my world any more than it had already. I wasn't being selfish or unconcerned about others; there was just a sense that until things changed, I should continue doing the only things I knew how to do. So when the school year ended, I worked through the summer of 1942, and started my sophomore year at KU, all the while, waiting to hear that I'd been called up to serve. Hundreds of Midwestern young men had already been drafted, and it became more and more obvious that few males—especially those of us with athletic body types and conventional attitudes of patriotism—would be excluded from the war effort. Deferments, even for older guys and those with physical problems or other mitigating circumstances, were getting fewer.

That school year, I played another season of football, and competed in track, basketball, and spring football practice again. A full year after the attack on Pearl Harbor, I was still in school. But it was getting to be an exercise in futility. Who could concentrate on school when almost every week somebody else close to you was going off to war? Each week, it seemed, we were having a farewell party for another buddy who was leaving. I never missed a single farewell party that I knew about. On the other hand, I missed about half my classes. I couldn't focus on my studies—as hard as I tried—and my grades suffered, plunging to C level, and threatening to dip even lower than that.

Dr. Woodruff, the academic dean, called me in for a conference in his office. We talked briefly about my languishing grades. He wasn't upset; he recognized what was happening in my life and the lives of so many other young men at KU, and he was a realist. "You're not making it in school, Bob. Maybe you just ought to enlist. Have you ever thought about joining the army?"

"Yeah," I said. "The thought just occurred to me." Actually, the thought had been in my mind for months.

My plans—like most everyone else's—were going to be changed. My younger brother, Kenny, had graduated from high school earlier that year, and our local draft board had already notified him to report. Could my notice be far behind? I weighed my options between enlisting or simply waiting to be drafted, and enlisting seemed to offer some better choices. There was a good chance that by joining the army's Enlisted Reserve Corps, I could at least finish my sophomore year, and possibly be assigned to the Army Medical Corps. Such a course might allow me to continue some portion of my education and, I hoped, enable me to pursue my dream of becoming a doctor when my service was completed.

On December 14, 1942, just before I went home for Christmas, I enlisted in the Army Reserve Corps. I was nineteen years old. The following June, I left the University of Kansas to join the world in war.

As I packed up my things, I couldn't help wondering if my college education was over, if I'd ever really come back, or if I'd ever get to play basketball for Phog Allen, or if I'd ever become a doctor. Of course, I intended to return to school as soon as possible. I hoped this war would be over quickly; that's what the rumors around campus were saying. But reality poked at my bubble of idealism. The First World War had dragged on from 1914 to 1918, with a lot of cleanup time after that. Nearly four long years. A lot could happen to a guy in three or four years. . . .

CHAPTER 7

Army Man

By December 1942, I'd been elected vice president of the Kappa Sigma fraternity, although it looked doubtful that I would serve very long. I'd soon be saying good-bye to the guys at KU. Yet, in some ways, I looked forward to going into the army; the anticipation and the uncertainties of not knowing were getting to me. Like someone facing an inevitable surgery, I felt I could put it off only so long. After a while you start thinking, **I just want to get it over with.** My restlessness seeped into my letters:

> **Dear Mom and Dad,**
> **Sunday night—9:30—I'm sitting here trying to study, but I can't, so I thought I'd write you a letter. . . .**
> **I haven't heard a word from the**

Army, so I'm still going to school, although I can't say that I'm getting much out of it. About every day, we hear a different story as to when we will be called, and I just can't see this studying when it's so uncertain. . . . My orders to report will come to Russell, so as soon as you get them, call me and let me know.

I'm running the quarter [mile] for KU next week in Kansas City in the big indoor track meet. I will run against boys from Missouri, Kansas State, Iowa State, Nebraska, and Oklahoma. I haven't much chance to win, but I'm sure going to try. The fastest anyone has run it indoors so far is 52 seconds and my best time is 54, but I'm not in very good condition.

My girl, Grace McCandless, is going to be initiated Saturday night, and I think it would be nice if I could send her some flowers or something, don't you, Mom? Her initiation will be over at 8:00 on Saturday, and then she's coming to K'City to watch the track meet.

Anxious as I was to receive some sense of direction from the army, I had a good reason for my reluctance to leave KU. Grace McCandless was a tall, dark-haired campus beauty queen, and one of the most charming young women I'd ever met. We became friends during the fall of my sophomore year, and before long we were dating.

Grace invited me to her sorority's spring formal, a sit-down dinner followed by a dance. Wearing my best and newest suit, I was a little nervous eating dinner with fifty girls, and I was more than a little afraid that I would do something wrong, as far as my table manners were concerned, but I managed. I just slid several of the smaller forks out of the way and used the largest. Grace didn't seem to mind, and we had a great time. A few weeks later, she invited me to her home in Hutchinson, Kansas, to meet her parents. That was more nerve-racking than the spring formal, but Grace's folks made me feel right at home.

Back on campus, our relationship continued to blossom, and by the time the spring flowers bloomed in Lawrence, I had asked Grace to wear my fraternity pin. "Getting pinned" was a step beyond "going steady," yet somewhat less than the giving or receiving of an engagement ring; it was tantamount to being pre-engaged, so before I did it, I discussed the matter with both Grace's mom and my parents. Everyone responded favorably.

I invited Grace home to meet my family—no small gesture in those days. It was one thing for a young lady to invite a young man home to meet the family. That was almost expected. But if a fellow took a gal home to meet his family, things were definitely getting serious. My family loved Grace, Grace loved them, and I was delighted. It didn't seem to bother Grace that my family was not well off, that our home had few fancy frills, or that my family members were common, straightforward, and unpretentious. Nobody ever accused the Doles of being part of Russell's well-oiled, well-to-do upper crust. We were just down-to-earth folks, just like Grace and her family, and we were happy that way. Grace and I seemed well matched.

Then, just about the time things were looking up for Grace and me, I received a notice from the army. Within a few days of arriving home from college, on June 1, 1943, I was to report for induction at Fort Leavenworth, about twenty miles northwest of Kansas City. I called Grace and told her the news, packed up my things, said good-bye to my family, and headed off to join the army.

I checked in—was "processed," in military jargon—and got my first dose of army life. After hurrying to Leavenworth to report, I waited . . . and waited . . . and waited. Along with the other inductees, I expected to be transferred to another camp to begin my basic training. I knew Mom and Dad would be anxious to know how I was doing, so on June 9, 1943, from Fort Leavenworth, I wrote my first letter as an active member of the United States Army.

Dear Folks,
  How is everyone? I'm feeling fine, and like it fairly well. Haven't left here yet, but expect to in a day or two. Probably going to Texas, although I don't know for sure.
  Has Larry [Gloria's newly married husband] gotten there yet? If he hasn't, don't forget to have him look up Grace when he goes through Hutchinson. I called her up tonight and practically spent a month's wages, $2.94.

I was on K.P. [kitchen patrol] duty sixteen hours Monday and I was really tired Monday night.

I made 130 on my test, so I qualified for Officer Candidates School, but I have to wait six months, I imagine, or at least until I finish Basic Training. After Basic Training, which lasts twelve to thirteen weeks, they may send me back to school. Just depends on how I do.

Has Kenny gotten his orders yet? When he comes to Leavenworth, tell him to wear comfortable clothes because you have to wait for four or five days, and a suit is too hot to wear. Also tell him to write Grace and have her send him those tests that she has. Have him study the one on mechanical aptitude and the ones on general intelligence, mainly. If you don't make a good grade on the intelligence test, you don't have as much chance; so it will pay him to study hard on the tests before he has to report.

We get up at 4:30 and go to bed at

**9:00 so I don't have much time to
myself.**

**It sure would be easy to get mad at
some of the sergeants, corporals, and
lieutenants around here, but it
doesn't pay, so one just has to grin
and bear it.**

Several weeks later, I was still stuck at Fort
Leavenworth. My assigned tasks included
checking men in and out of the barracks, conducting bed checks, and running errands for the
corporal in charge—pretty soft work compared
with what most of the new inductees were doing. But I was anxious to get with the "real"
army program, even though I'd heard that basic
training could be rather grueling—football practice to the max.

Mom, Dad, and my friend Phil Ruppenthal
visited Grace's family in Hutchinson, and I was
confident that they'd be as taken with the Mc-
Candless family as I was. At the end of June, I
wrote Mom, "What do you and Dad think of
[my] buying Grace a ring if I have enough
money saved? Don't you think she's worth it?" I
had no idea how long it would take me to save
up enough money to afford an engagement ring,

but I knew that I didn't want to let Grace slip away.

Whether for security reasons or on account of the massive amount of clerical work necessary to keep track of everyone in the military, the army was notoriously closemouthed about troop transfers. Consequently, although I knew I'd be going somewhere soon, I was far from sure where the army would be sending me for basic training. "I finally found out when I'm leaving this place," I wrote home on July 5. "I get my shipping orders this Friday, leave Saturday, and reach my destination Sunday or Monday. I don't know for sure where they will send me, but I have a hunch it will be Camp Barkley, Texas."

Meanwhile, just as I was getting ready to leave for basic training, Kenny was called to Leavenworth, along with several other boys from Russell. I got to see them before I left, and Kenny delivered a package of "goodies" from home. I wrote Mom and Dad again on July 8:

> **Saw Kenny about two hours ago, and got all the stuff that you sent down with him . . . I'm still not sure whether or not I'll be here this weekend, but I will know tomorrow**

**night about six o'clock. If I am here, I
will try to get a weekend pass, and
come home. I'm getting hungry for
some fried chicken and some
homemade ice cream. They don't give
us much milk in the Army, and, as
well as I like milk, I really miss it. The
food as a whole is pretty good. We
don't have to worry about rationing
anyway.**

Two days later, on July 10, I received my orders to move out. I still didn't know where I was headed, but I wanted to let Mom and Dad know that I wasn't in Kansas anymore. And I probably wouldn't be, not for a long time. "By the time you get this," I wrote, "I will be somewhere on a train, though I don't know where it will be. I got my orders last night, and I'm leaving this afternoon at 3:30. I haven't any idea where I'm going, but I'm sure glad to get out of this place, and get started on something. I would like to have come home this weekend, but there isn't any chance of that now.

"Kenny is in the same company that I am here, and I'm trying to get him a pretty easy job. . . . He looks nicer in his uniform than I did, and I think he is going to like the Army fine. He's with the rest of the Russell boys, and I imagine they will all be shipped out at the same time, so they should get along all right. . . . I guess I will have to take a rain check on the fried chicken and ice cream, but I will be home plenty soon enough, I guess."

My hunch proved correct. The next morning I woke up in Camp Barkley, an army training installation about eleven miles southwest of Abilene, Texas. The camp was enormous, with row after row of buildings and barracks stretching across more than seventy thousand acres of land. At its peak, Camp Barkley housed more than fifty thousand soldiers. The first thing I noticed when I stepped outside—besides the wilting heat—was the barrenness of the land around the barracks. There wasn't a tree anywhere in sight. I was assigned to a barracks hut with fourteen men in one room, and no air conditioning.

I was to be a member of the Army Medical Corps, which was all right with me. I hoped that maybe I could learn some things in the military that would help me later in my career as a doctor. In reality, it didn't matter much right now. Basic training was just that—basic—and it was pretty much the same for everyone in the army. On July 13, writing on army-issued paper complete with soldiers and flag insignia on the header, I passed along to Mom and Dad my first impressions of my new home, and a rough idea of the army's plan for my new comrades and me:

They say our basic training will be plenty rough, so I think that I'll like it. . . . Our non-coms [non-commissioned officers] are pretty good fellows, though they act pretty tough, and swear quite a little. I'm a squad leader in my platoon; Dad should know what that means. Basic Training lasts eleven weeks so it should end sometime in October. There are three hundred men in our company. Of these, about eighty will go overseas, four to Officer Candidates School, ten or fifteen to school, and the rest will be sent to some base hospital for technical training.

I added more in a letter the following day.

They issued us our gas masks today, also our tents and field packs. They also made us all get crew cuts, but don't ever say anything to Grace about it. They cut everything off but our ears, so you can imagine how I look.

I'm anxious to start basic training and get it over with so that I can go either to school or O.C.S. They are talking about lengthening our basic training from eleven weeks to sixteen weeks. I hope they don't, but it doesn't do any good to worry about it. Just finished writing Grace a letter, told her about everything but my haircut; by the time I see her again, it will be down to my knees probably.

Old soldiers say that the military makes you appreciate the good things in life, and after only a few weeks in the service, I was ready to agree. I could already tell that I had a fresh perspective on how good we'd had it growing up at home in Russell. Writing to my sister Gloria and her husband, Larry, as well as Norma Jean, who was staying with them in New Mexico during Gloria's pregnancy, I expressed a bit of nostalgia:

> I try to write the folks every day, for I imagine they are pretty lonesome with all four of us away from home. It should give Mom a little rest with none of us there to clean up after or to cook for. I sure realize how swell our parents are to us, and I know that Kenny, Norma Jean, and you feel the same. We've never gone without anything that we really needed; and when I see some of the fellows around here, I appreciate the fact that the folks did watch us pretty close. Try to write them as often as possible, for I know they hate the thought that we are all grown up and away from home.
>
> If you kids haven't time to write to both Kenny and me, write to him first, for I'm used to being away from home and it doesn't bother me too much, but this is Kenny's first time away from home and I imagine that he is pretty homesick about now.

I had gotten in the habit of writing home regularly, and I wasn't sure how much time I'd have for writing once basic training began, so later that day I scribbled another letter to Mom and Dad.

Today is Sunday and our day of rest in the Army. We got to sleep in until seven, then we had roll call. After that, we were released to do as we pleased. I have plenty to do, so I won't have much time to play around. I have some clothes to wash, shoes to shine, and quite a few letters to write. Tomorrow our Basic Training starts, and if it is as tough as everyone says it will be, I probably won't have much time to write. . . . Our Basic Training starts with a lot of drilling and exercises. After a couple of weeks, we start going to class to study First Aid, Poison Gas, Stretcher-bearing, and everything that a medical soldier should know. The only thing that I don't like about it is that a medic never carries a gun, and they don't even train us how to use one. The only

protection that we would have overseas is a medical red cross on our sleeve. I'd hate to think of being shot at and not being able to shoot back.

I should have a pretty fair chance of either going to school or to Officer Candidates School. Out of the 250 in our company, they pick about four for O.C.S. If I did get a chance to go, I don't know whether I would or not. Grace wants me to go to school, and I imagine that you both do, too. They don't always give you your choice in the Army, though, so I'll do whatever they tell me. I would like to be an officer, but I talked to a Second Lieutenant the first day I was here, and he told me that if he had the chance to go back to school, that would be the best thing to do.

The soldiers who had warned us about the rigors of basic training weren't exaggerating. It was tough, but I didn't mind. The sergeants kept us plenty busy, though. As if we were training for an athletic team, our days were filled from morning to evening, from reveille to taps, with conditioning exercises, drills, and classes. Military life reminded me of the camaraderie I had enjoyed back at KU with the guys in the fraternity and on the football field and basketball courts. But as our drill sergeants constantly reminded us, we were training for real; this was war, and on the battlefield, teamwork, obeying commands, and the discipline of the guys in uniform could mean the difference between life and death.

On July 22, 1943, while still in basic training at Camp Barkley, I turned twenty years old. It seemed strange having a birthday without any family members around to celebrate. But I did receive a great gift. I learned that Kenny had been sent to Camp Barkley as well, and we'd soon have an opportunity to be together. Nevertheless, I was a little down as I wrote my folks that night.

Dear Mom and Dad,
  Spent a quiet, but not too happy birthday today. I guess I missed my banana cake and all the ice cream and different stuff I'm used to having. I'm twenty years old now, and about time that I was starting to think a little more and talk a little less.

Tomorrow I have K.P. so I'll be plenty busy. I suppose you know that Kenny is only about a mile from where I am. I haven't seen him yet, but I'll try to get away Sunday long enough to see him. Seems kind of funny; Kenny and I are down here during the Second World War, and Dad, you were in Texas during the last war. We will have to have a big reunion when the war is over. . . .

Took our first hike yesterday, five miles with a pack, gas mask, and pistol belt. Sure did get hot. About 110 degrees yesterday.

Going to bed now, for I have to get up at five in the morning for good old K.P.

Love,
Bob

A few days later, I got another great surprise—a letter from my dad. As I mentioned previously, Dad wasn't one for a lot of words, but now that both of his boys were in the army, and both of his daughters were several states away, he decided to write to us. In his letter,

Dad hinted that Mom wanted to come see Kenny and me. Mom enjoyed traveling, and wouldn't have had the slightest hesitation about getting on a train by herself and making the trip to Texas to see us. But I didn't think that was such a good idea.

I wrote back:

Dear Folks,
  I saw Kenny last night and all afternoon today. He didn't seem too homesick to me, so he might be getting over it by now. I ate some of the cake you sent him, and it was really good. This afternoon, Kenny, Melvin, Ernie, and I went to the service club and ate ice cream all afternoon. We were going to call you, but the long distance operator told us we would have to wait four hours, so we gave it up.
  Dad, you wrote in your letter that you wanted Mom to come down and visit us. I really don't think that it would do her much good to try it, for we never know what we will be doing. We have about two or three hikes a

week starting next week and part of
them will be at night, so I really doubt
that we could see her much unless it
would be on Sunday. After four or five
more weeks, it might be different, and
if I do think that we could arrange
something, I'll write you to let you
know. . . .

I really appreciated your letter, Dad;
I think that is the first one I've ever
had from you. I read the one you
wrote Kenny, and everything sounded
swell.

Had a letter from Norma Jean,
Thursday. I guess she is really having
fun.

Goodnight Mom and Dad
   Love,
   Bob
P.S. I don't need a thing—do you?

Training intensified the following week, including an all-day, twenty-mile hike. Besides
that, I was moved to a new battalion in another
part of camp. The good news was that when the
moving dust cleared, and I was settled in at
another barracks, I was only about one hundred

yards from my brother, Kenny. He and I went to the service club and celebrated by eating practically all the ice cream they had in stock.

That same week, I informed my parents of my first army payday. "I got $74.25 for two months. I had about $25.00 already, so I'll send $80 home. . . . I'm not going to take a chance on sending the money in an envelope, so I'll send it Saturday by money order. I'd hate to lose it, for it was the hardest money I've ever earned."

I wasn't exaggerating; the ground drills were arduous. Beyond that, although I recognized the importance of knowing how to do such things, crawling in and out of foxholes and ducking machine gun fire seemed like an extremely remote possibility for me. It was hard to imagine that I'd ever have a need to use those skills to save some soldier's life, let alone my own.

By now, the United States was a nation mobilized for war. Everything from gas to food was carefully rationed by the government. The entire country was affected by the war effort. Nevertheless, like many moms and dads of soldiers, my parents continued to send Kenny and me "goodie boxes" as often as possible. Ours just seemed to have more goodies than many of the other guys'. When Kenny or I would receive a package, our barracks would mysteriously fill up with guys hoping to get a taste of candy, cake, fruit, or some other treat from Bina and Doran Dole. I knew that my parents didn't have much to spare themselves, yet Mom and Dad amazed me with their generosity. In my next letter home, I tried to convey my thanks:

The box I got today was so big that I could hardly carry it. Kenny hasn't been over to get his stuff yet, but he probably will before I get this written.

I've had a sore foot the last three days, so this morning I had it lanced. It was a callous which had become infected on the bottom of my foot, and the marching didn't help it any. One of the trainees in our company was a foot specialist before he was drafted, so he fixed it for me this morning for forty-five cents. . . .

I have a question to ask you and Dad. Would you rather I go back to school or try to go to Officer's Training? I have to make up my mind

pretty soon, and I would like to know which you would rather I do. If I go to school, they think that I would have to stay five years after the war is over. There are about seventy boys in our company eligible for Army Specialized Training, and some of them, like me, aren't sure whether they want to stay in the Army five years or not.

I would like to go to school, and I would like to be an officer, too. I can't do both, so I have to decide within the next week or two.

To go to O.C.S. one has to complete basic training with a recommendation from the commanding officer. I would also have to be interviewed five different times by high ranking officers. If they thought that I had the qualifications, and I was capable of being an officer, I would then go one month to Officer Candidate Preparatory School. If I got through that, I would then go four months to O.C.S. and, providing I passed, would graduate as a 2nd Lieutenant in the

Medical Administration Corps. It's a long, tough course, but it may be worth it. . . . What do you and Dad think I should do?

Before I close, I want to thank you for the food and shoes. The shoes seem a little big. I haven't worn them yet, so if they are, I can send them back, I imagine. I like the style all right, though. The candy and cookies are really good, and oranges are my favorite fruit.

You needn't send me stuff so often, for I imagine it costs quite a little, and it probably takes most of your food coupons. . . . Well, good night folks, and thanks again for everything.

Love,
Bob

The following week, I received another box loaded with even more goodies.

About the time we were half done with basic training, we received the news that basic had been extended to seventeen weeks instead of eleven or twelve. It really poked a hole in my plans. I had already decided that I wanted to buy Grace an engagement ring on my first furlough. Now my plans had to be delayed. Trying to look on the bright side, I figured, **Well, at least I'll be able to save up a little more money.**

A few days later the news got worse. "Today, they gave us our steel helmets," I informed my folks. "They wouldn't let us go to church this morning, for we had to wait until they had all the helmets checked out. I guess we start practicing the machine gun course pretty soon. It isn't dangerous, and they won't start using real bullets for three or four weeks. They generally don't issue helmets until the last week of basic, but I guess they decided that since our basic training is now twenty-one weeks instead of twelve, we may as well get used to them. They told us for sure that our basic training was to be twenty-one weeks; first it was twelve, then seventeen, and now twenty-one. The war will probably be over before I even get out of this place."

Having to remain in basic training almost twice as long as I had anticipated was difficult to take. Breaking the news to Grace was even harder. She surprised me, though. Rather than being despondent about the delay, she decided to close the distance between us. She suggested that she come to Texas to visit me. As much as I longed to see her, I thought a visit to Barkley was a bad idea, and I wrote telling her so. I conveyed the same message to my mom.

Grace wrote to me three days ago and she wants to come down between the 6th and the 13th of September. School starts the 19th so she would have to come then; however, I wrote to her that I would write and see what you thought about it. I don't think it would be very wise to try and make it. In her letter today, she said she was going to write you and try to talk you into it. I just finished writing her, and in the letter, I advised her to stay home and get ready for school. I don't know where you would stay, and I could see you only about five hours a day. Then if we would have any work at night, I couldn't even see you that much.

You think it over. It would cost about $30.00 or $40.00 for a round-trip ticket on the train, and I doubt that you could get enough gas for the car.

You may be able to work something out, but I know she couldn't afford to come on the train. I'll be home before long, so there's no use spending any money to come down to this hot place. You can tell Grace that there isn't much chance of trying to come down. It may sound as though I'm trying to discourage you, but I'd rather you'd save the money and buy something you and Dad need, or put it in the bank for Kenny and [me].

In one sense, I was extremely glad that Grace wanted to come down to Camp Barkley to see me—**She must really love me**, I thought, **to be willing to make such a long trip**, but I kept stalling her nonetheless. I threw out every roadblock I could think of—it's too hot, too dirty, it would cost too much to travel all the way from Kansas to Texas by train. Actually, I just didn't want Grace to come because the last time she had seen me, I had dark, wavy hair. But when I was inducted and shipped out for basic training, like millions of other soldiers, I sat down in the army barber's chair and heard the buzzing sounds of the military shears, as the

Love and sacrifice shaped my life.

Top: Mom and Dad, Bina and Doran Dole. Mom sold sewing machines to help make ends meet. Dad missed only one day of work in more than forty years. **Center:** The children of Doran and Bina Dole **(clockwise from top left):** Kenny, Bob, Norma Jean, and Gloria. **Above:** The home in Russell, Kansas, where I was born and spent my early years.

The "greatest generation."

Top left: My younger brother, Kenny, served in the Pacific during the war. These photos were taken during Mom's surprise visit to Camp Barkley, where she told the army officer that he should do a better job of taking care of her boys. **Above:** It's said often that my generation is the "greatest generation." That's not a title we claimed for ourselves. Every generation of young men and women who dares to face the realities of war—they are the greatest generation.

The war touched us all, one person at a time.

Top: At Breckenridge, Kentucky, I trained with anti-tank gunnery. I worked out regularly to keep in shape so I could compete in army track meets. **Center:** With one of my good friends from Russell, Bub Dawson **(right)**. As a teenager, I worked as a soda jerk at the drugstore owned by the Dawson family. The war separated us, but later, along with my buddy Phil Ruppenthal, Bub and Chet Dawson helped raise funds to pay the hospital fees for my operations after the war. **Above:** Tossing Larry Jr., Gloria and Larry's little boy, in the air with ease.

A lot of blood was spilled on both sides.

Top: The Second Platoon—my guys. These men fought valiantly on Hill 913. Many would not return home. Sergeant "Ollie" Manninen, who dragged me out of the machine-gun fire, is in the top row, third from left. Sergeant Frank Carafa (lower left, without a helmet), pulled me to safety. Private Arthur McBryar (standing, fifth from right) stayed with me for more than six hours till help came. I'm kneeling at the far left, in a light jacket, without a helmet. **Above:** Castel d'Aiano, the mountain town where I was wounded. It was occupied by the Germans in early 1945 and received a horrific pounding by Allied bombers.

Stuff was coming at us from every direction.

(Courtesy of Denver Public Library)

(Courtesy of Denver Public Library)

Top: It was a scene similar to this where radioman Eddie Sims went down. Seeing that he'd been hit, I crawled out to help him. I grabbed Sims by his shirt and started dragging him back toward a shell hole. That's when I felt the sting in my right shoulder. **Center:** This is not me, but the men who transferred me from Hill 913 to an army aid station had to carry me down the rugged, bombed-out Italian hillside on a stretcher much like this. **Above:** The hospital in Pistoia, Italy, where I was taken shortly after being wounded. This picture was taken during a recent visit, in November 2004.

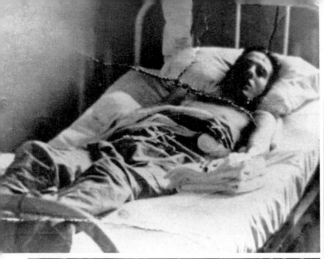

On the road to recovery.

Top: After being flown to Miami packed in a plaster body cast, I was shipped to Winter General Hospital in Topeka. The doctors gave me little hope of surviving. **Center:** I'd lost a kidney, had frequent tremors and spasms, and still could not move my arms, but was determined to walk again. **Above:** I was looking for a miracle when I went to Percy Jones Army Hospital in Battle Creek, Michigan. What I found was a group of fellow soldiers who were all trying to rebuild their lives. One of the life-long friendships I formed at Percy was with Danny Inouye **(far left),** who was wounded in Italy one week after I was and only a short distance away.

We never walk alone.

Top: With my sisters, Gloria **(left)** and Norma Jean **(right)** at our parents' home in Russell. Nearly two years after I'd been wounded in the war, my arm had atrophied and I'd lost seventy-two pounds. Despite seven operations and the valiant efforts by Dr. Hampar Kelikian to make the right arm functional, it improved very little. **Center:** At home in Russell, 1947, using the exercise pulley Dad, Kenny, and some friends made from lead window-sash weights. **Above:** With Dr. Kelikian, 1965. Dr. K refused to take any money from me for the complicated surgeries that he performed on my right arm. He taught me to focus on what I had left, rather than on what I had lost.

A tribute long overdue.

Top: With General Dwight Eisenhower, our leader during World War II, a Kansan. I admired Ike as a commander, and even more as our president. **Center:** It took more than eleven years, from inception to completion, but at long last we had a fitting memorial to honor the American men and women who endured World War II. On the right is Ray Kaskey, the principal sculptor on the project, along with several members of his crew. **Above:** At the National World War II Memorial on the Mall in Washington, D.C., dedicated on May 29, 2004, in honor of the 16 million Americans who served in the armed forces during the war.

barber shaved my hair off. I looked up in a minute or so, and realized I was nearly bald. "Next!" the barber called.

I had a bit of stubble on my head, but not much more. Call it foolish, stubborn pride. I didn't like the way I looked and I didn't want Grace to see me like that, so I stalled her coming to visit, grabbing at any excuse I could find. I didn't really need to make up excuses. Camp Barkley was not exactly a tourist site. Intensely hot and arid in late summer, dusty, with no good place nearby for visiting guests to stay, Camp Barkley was as close to hell as any human would want to get. Still, Grace insisted on visiting before the next term at KU got into full swing, and my reluctance in letting her come strained our relationship.

On September 3, 1943, I wrote to my folks: "Grace seems to be pretty disgusted because I didn't want her to come down. She'll get over it in a week or two, and I still don't think that she should come." Later that month, I wrote to my dad:

This Army life is okay. I get tired of taking orders, but it is really going to be good for me later on. I'm the

leader of my platoon; I'm right guide, if you remember what that is. The sergeant likes me pretty well, so I'm getting along pretty well.

I'd like to be home, but there are seven million other boys who would like to be, too! They keep us pretty busy, and you'd be surprised how fast time passes. I don't get too homesick, for being away at school for two years helped me. Kenny seems sort of blue occasionally, but he'll get over it before too long. I think he deserves more attention than I do until he gets his spirits up. I'd try to write him every day, for he always seems to feel better when he gets letters. Have all the relatives write to him if they will.

I'm glad you put $25.00 in the bank for me; I'm going to try to add $5.00 to it every month so I can buy Grace a ring before too long. I really miss her, but there's not much I can do about it. I guess I should have had her come down, but you should know how conditions are around an Army camp. No places to stay, and all that kind of stuff. I may get to see her in November or December for I'll be eligible for a furlough then. If I go to O.C.S., I won't get out until April or May.

Well, Dad, you and Mom take it easy and don't worry about us. This place is just like a Boy Scout camp.

Tomorrow is Sunday, so we're going to church. "Moon" is going, too. He's a little short on change, and he thinks he can get a little out of the collection plate.

I continued to stonewall Grace about her proposed visit, and I actually breathed a sigh of relief when the new school term began in September. But if I thought that I had dissuaded my mom from coming, too, well, I should have known better. One day, Mom simply showed up at the base. She had bought a train ticket and made the trip by herself, all the way from Russell to Abilene to check on her boys, and to make sure the U.S. Army was treating us well.

Now I can't attest to its veracity, but the way I heard the story, she talked her way past the sentry at Camp Barkley and literally started hiking down the dusty main street, looking for Kenny or me. Keep in mind, Camp Barkley was bigger than our hometown of Russell, yet Mom fully expected to bump into one of us.

Fortunately, the military police corralled Mom before she got very far. They tried to talk her into leaving, but they didn't know Bina Dole. "I have two boys here," she told them, "and I've come to visit them."

"Ma'am, this is an army base," they said. "You can't just go wandering around here." They called a first sergeant to deal with her. When the noncom showed up, Mom chewed him out. Apparently she told him that he should do a better job taking care of her boys. (Every concerned military mom or wife in American history has probably known that feeling.)

I'd hate even to think what the sergeant thought about Mom's remarks. **Sure, lady. You sent me a couple of pansies down here, and now you want me to baby them!** Nevertheless, the first sergeant politely told our mom, "Wait here, ma'am, and we'll see if we can find your boys for you."

He found us, and we actually got to visit with Mom for a while. For years after that, one of the classic stories around the Dole dinner table was the time Mom came to Texas and chewed out the army sergeant.

We were soon glad that Mom had made the effort. By the end of the year, Kenny had received orders assigning him to the South Pacific, and I was transferred to the Army Specialized Training Program (ASTP) at Brooklyn College, in New York. It would be a long time before we were together again. By then the war would have changed all our lives.

CHAPTER 8

Training for War

It has long been assumed that age bestows wisdom. Not necessarily. The only thing that automatically comes with age is wrinkles. Still, with enough experience, one gains a certain perspective. Perhaps one of the great benefits of growing older is that you have a longer vantage point from which to look back on your life, noticing the milestones and appreciating the pivotal moments, when seemingly insignificant decisions were made that affected the rest of your days. Sometimes our great disappointments can become fresh opportunities. Such was my decision regarding Officer Candidate School.

I was one of twenty-seven young men from our company who had applied for OCS. We tested, interviewed, and drilled for two majors, two first lieutenants, and our company commander. Then we took turns drilling one another. I was encouraged, especially when ten of the other guys withdrew their applications after the first round of tests. I felt that my chances were good, but still not great. I wasn't too worried about it, though. I figured if I didn't get into OCS, then it would be a sign for me to apply for the Army Specialized Training Corps; in my case, the choice between becoming an officer and going to school while in the army would be decided by default.

During my first few months in the army my weight had dropped, partly due to the change in diet, no doubt, and partly because I was stationed at Camp Barkley. Who feels like eating that much when it's 110 degrees in the shade? (If there had been shade.) But by late September I had grown accustomed to army food, and was back up to 183 pounds. I wanted to beef up a bit, because we were scheduled to go on a three-week bivouac—training maneuvers during which we would be sleeping in tents, eating nothing but K rations, and living out in "the wilds"—at the end of October.

I hadn't heard anything one way or the other regarding OCS, so I wrote to Mom and Dad to let them know I was still living in limbo. My problems were small compared with theirs. The war was tightening everybody's belts on the home front. Gas was getting more and more scarce; even when food-ration stamps were plentiful, there was not always food available in the stores. Most people, our folks included, had some sort of garden in which they could grow a few vegetables. That helped, but not much. Mom and Dad never complained. Knowing their economic problems, I tried to be upbeat in my letters: "How is everything at home?" I asked with feigned casualness. "I just heard that they were cutting gas down to two gallons a week instead of three. I guess you two will have to start walking now. It's been so long since I've ridden in a car that I don't think I would know how to act. I've been feeling good for a long time now. I guess the Army life must agree with me. Hope you two are feeling okay."

I was still ambivalent about the choice between OCS and ASTP when I wrote to my folks on October 14, 1943.

We had a twenty-five mile hike yesterday, but fortunately, I didn't have to go. I had an O.C.S. interview yesterday morning, so I was pretty

lucky not having to go on that jaunt. I also had an A.S.T.P. interview Tuesday morning, and if I don't get O.C.S., I'll have a chance at that. I really am not planning on O.C.S. for they are going to have only one more class which starts the 19th of November, and that will probably be filled with boys from overseas. The classes are only 250 men and there are already 500 on the waiting list. I've done the best I could, so I am not going to feel too bad if I am not selected. If I do make it, I hope that you and Dad won't think that I did the wrong thing. . . .

I haven't heard from Grace much lately. I don't know what's wrong, but I'll probably know before too long.

Take it easy folks; don't forget to save us a turkey for Thanksgiving.

Love,
Bob

Don't need a thing but sleep.

A few days later I talked to my company commander, and he gave me some good news. "As far as I know," he said, "you are the only one of the twenty-seven applicants who made OCS. Congratulations, Private Dole." I was pleasantly surprised. I didn't think that I had a chance.

That week, I left for the seventeen-day bivouac with all sorts of questions in my mind. With only three days of basic training remaining when we returned to camp, I wondered, what was next on the army's agenda for me? Where would I be sent? Would I make it out of OCS before the war was over?

Closer to my heart, I sensed that Grace's feelings for me had changed. I rarely heard from her anymore, and when I did, there was never a hint about our future together. I figured that our relationship was in trouble. For the next seventeen days, however, there was nothing I could do but camp out and wait.

When I returned from the two and a half weeks of maneuvers, I found a message informing me that, just as the commander had said, I had been the only applicant from our company to make OCS. Unfortunately, the quotas for entering were full. I could wait five or six months for another opening, or apply to ASTP. I decided on the latter. I was sorely disappointed, but there was no use bellyaching about it.

Along with the OCS letter, there was also a small package from Grace waiting for me. Expecting a gift, or some token of our love, I hurried to the barracks and sat down on my bunk to open the package in private, even though I was still filthy from the rigors of the bivouac. Most of the other guys were opening their mail or already heading to the showers. Nobody noticed as I quickly peeled the brown shipping paper off the tiny package and found a small box. I tugged the lid off the box . . . and the contents brought me up short.

There, tucked into a small tuft of soft, white cotton, was my fraternity pin.

For a moment I had to fight back the tears, as I realized the significance of Grace's returning my pin to me. Along with the package, there was also a letter from her, explaining that she had met a fine young cadet who was stationed at the naval air base in her hometown of Hutchinson. Our friendship would continue, but any possibility of marriage had gone out the door.

It was only midday, but I felt as though a dark cloud had descended on me. For the next few hours, I simply went through the motions. I pulled off my dirty clothes and stumbled toward the latrine and shower area. I slowly shaved my heavy growth of beard and then stepped into the hot shower. I kept my face under the showerhead for a long time, partly because it felt so good to be clean after being filthy for two and a half weeks, and partly to conceal the tears.

That night, as I had most nights since joining the army, I wrote to my folks. For some reason, I felt compelled to write two separate letters, which I had done on occasion before, but not frequently. "At last, bivouac is over after seventeen days," I told my mother. "We got back at 11:30 this morning and by the time I took a bath and shaved, it was about one. I'd never been so dirty before in my life, and the shower really felt good.

"Just wrote to Dad, so now it's your turn. I'm going back to school, if I can, so I may end up back at KU, if I'm lucky. I don't know how long I'll be here before shipping out, possibly a week or two, or maybe even longer. I was accepted for OCS but their quotas are full, so rather than wait four or five months for an opening, I'm going to try ASTP, which is probably better than OCS in many respects anyway. . . .

"There isn't much to write about, so I can't write too much. My basic ends this Saturday, and then, perhaps, a furlough, but I doubt it very much.

"You needn't buy Grace anything, Mom. I think she has found herself another man. I'll send you the pin she was wearing, so don't lose it. I'll send it in a day or two. When you wear it, watch the clip on it, for it doesn't catch very well. I guess I'll have to find me another girl somewhere. Tell Norma Jean to look around for me. I'd rather have a blonde, if she could find one. I felt pretty bad when I got the pin back, but I'll get over it in a month or two."

That same night, I wrote to my dad of my experiences on maneuvers, and also about the possibility that the army might send me back to KU to study. But of Grace I said not a word. Instead, I kept Dad's letter more lighthearted. "Well, basic is almost over with bivouac completed. . . . I'm sending you a box of rations. Open them up and see if you can make a meal out of the contents. We lived three days on K rations, and they aren't very good. If you like them, you'll be the first person I know that does. . . .

"Write if you have time. Might be home in a week or two, but don't count on it too much. So long, Bob."

• • •

For the next week or so, I moped around the camp, feeling sorry for myself. Making matters worse, everyone in our company received a furlough except the ASTP guys—and that now included me. The barracks seemed dead with everyone gone. I had plenty of time on my hands, so I indulged in a bit of self-pity as I wrote to Mom. "I am supposed to leave here tomorrow or Friday for Camp Maxey in Paris, Texas, where I will take several tests. If I pass these, I will be sent to school to study medicine, language, or engineering. I'm not very optimistic, for they say that fifty percent of the candidates fail the tests. . . .

"My letter writing has been cut down quite a little now that I haven't a girl to write to. You wear my pin, Mom, for I'm through with girls until this war is completely over."

I enjoyed the trip to Camp Maxey, and apparently I passed the tests with flying colors, because the next thing I knew, the army was shipping me out to Brooklyn College, in New York, to study engineering. Engineering? I wanted to be a doctor. But as I was learning, the army wasn't always concerned about what I wanted. I was certain that somebody wearing stripes could tell me why a man in the medical corps needed to study engineering, but for the life of me I couldn't explain it. Moreover, the army offered no such explanations. "We need bodies in engineering, Dole! You're now in engineering." On the plus side, if I made it through the intensive studies, I might get into the war before it was over. My enthusiasm spilled over in a letter dated December 6, 1943, as I informed Mom and Dad of my new assignment.

Dear Folks,
  Well, I'm leaving for Brooklyn College in Brooklyn, New York tomorrow morning. I've always wanted to go to New York, so I guess I should be satisfied. I'd rather be closer to home, but there is nothing I can do about it now.
  I guess the school is really nice. All the buildings are practically new, from what I hear, and the girls outnumber the men six to one.

The odds convinced me to rethink my pledge.

The other ASTP guys and I traveled together by train from Texas to New York. The trip took two and a half days, as we passed through half a continent, most of which I was seeing for the first time. I was fascinated looking out the windows of the train, seeing our beautiful country. Even coming from the open plains of Kansas, I had never realized the United States was so large. We had quite a trip, with the only drawback being that we couldn't get Pullman accommodations, so we had to sleep in hard-backed seats.

Brooklyn College was barely twenty minutes from the heart of Times Square, and I could hardly wait to explore the place.

I checked out my books—thirteen in all. I'd be taking courses in physics, English, algebra, trigonometry, speech, chemistry, history, and the Constitution. I had all those classes in one term, plus physical conditioning and military drills. The academic program was brutal. Each week, I was scheduled to attend class for twenty-four hours, study twenty-four hours, and engage in five hours of physical conditioning and five hours of drills—every week. The army planned on getting its money's worth out of my education.

Brooklyn was an education in itself for a kid from a small town in Kansas. Back in Russell, most of the people were of German or other Teutonic descent. We barely knew anyone from an African, Asian, or Hispanic background. But New York was truly a melting pot of the world's cultures. The sights, sounds, and smells of the city amazed me. Just walking down the street or riding the subway was a cultural experience— speaking of the subway, we sure didn't have that in Russell! I heard people speaking all sorts of languages, some of which I recognized as English. But the English in New York was spoken with such a diversity of accents, I sometimes had difficulty understanding my own language.

If I was awestruck at the sights and sounds of New York City, I was nonetheless surprised at how little many New Yorkers knew about Kansas. I couldn't help poking a bit of fun at some of my new acquaintances. I told my parents, "These New York people still think there are buffalo and Indians in Kansas, and they think the only city there is Kansas City. They believe anything that you tell them, so some of the Kansas boys have a lot of fun teasing the girls around here."

My studies were more difficult than what I'd

experienced at the University of Kansas. I said
as much to Mom and Dad about a week before
Christmas:

> **I spent more time in school here this
> week than I did in a month at KU.
> They throw assignments at us so fast
> that we have to take our books to bed
> with us to keep up. I've already had
> seven tests and will probably have
> more this week. Study hall starts at 8
> and lasts till 10 PM. We have from 10
> till 10:20 to get to bed; then we get up
> again at 6:00 AM to eat and start
> classes. We have off from 3:30
> Saturday afternoon until 5:00 Sunday
> afternoon. I spent this weekend
> sleeping, but I'll probably spend the
> rest of them studying if I expect to be
> here very long.**

Despite all the work, I was enjoying my stay
in New York. The city was bustling with Christmas shoppers, but I had trouble immersing myself in the festive spirit. Besides the fact that I
had no money to buy presents, it bothered me

being separated from my family during the holidays, so far away, with the world at war, and with so many of our friends in the heat of battle somewhere. "Christmas is only a week away," I lamented to my folks, "but it sure doesn't seem like Christmas for some reason. . . ."

I spent Christmas day with some of the other guys in the military program at Brooklyn College, and we were all homesick for our family and our traditions. But I wasn't without presents. Mom and Dad sent a box of Christmas treats, along with a new sweater, some candy, and a fountain pen, one that I had really wanted. Norma Jean and Gloria sent socks, ties, handkerchiefs, and some sweet-smelling soap. From my brother-in-law Larry came a new shaving kit. It was especially thoughtful and self-denying of everyone to go out of their way to send Christmas presents to me. They did something similar for Kenny. I guessed that they'd spent all their ration stamps on us.

On New Year's Day, I wrote Mom and Dad to thank them for the presents and to bring them up-to-date on my progress.

The candy was real good, though I'm afraid that I got a stomach ache from eating too much of it. I opened my box up and everyone in the room made a dive for it! Everything was eaten except the paper, so you can imagine how they liked it. I had my [new] sweater on a while ago, and I'm afraid it's a little small for my 190 pounds. It might just be my imagination though, for I haven't had a sweater on in seven months. . . .

My Christmas presents all came about the same time, and the only thing more I could have asked for was a fifteen day furlough so that I could have been home with you and Dad. Which reminds me, Dad, that $20 you sent me arrived just as I was leaving last weekend, and it really came in handy. They gave us two and a half days off for Christmas, so the $20 is scattered all over New York by now. . . .

I had a letter from Lt. [Bub] Dawson and he plans on seeing me as soon as he has the chance. I guess I'll trade him uniforms or borrow one of his and see how it is to have people saluting me for a change. . . .

If you hear from Grace, let me know what she has to say. I still haven't been able to forget her, for she was such a nice girl. There are hundreds of girls around here, but they don't seem to interest me very much. . . . Happy New Year, Mom and Dad. I thought up a slogan, something like the one Ernie and Dad used to dream up every New Year's Eve. "Let's end the War in '44."

Dad's probably got a better one, but this one isn't too bad for a child of twenty.

My classes at Brooklyn College seemed to be getting tougher every day. We finished a full eighteen-week course in algebra in five weeks. The sheer intensity of the program was overwhelming. About 20 percent of the guys were flunking by the first of the year. I hadn't failed any tests yet, but I was working harder than ever before, and still having trouble keeping up. Chemistry and physics were my two most challenging subjects. Aside from all my school work, I was now the cadet first sergeant of "B" Company. There were two hundred men in my company, and my duties included filling out the sick book, taking charge of evening study hall, making announcements, and doing other odd jobs. It wasn't hard work, just time-consuming, and time was a precious commodity now.

Toward the end of the month, I received a surprise phone call from Bill Bunt, one of my fraternity buddies from KU. He told me he had joined the navy, was stationed in New York, and was attending midshipman's school. He'd been in New York for only three months, and in just two more months he would graduate as an ensign in the navy. Two other guys from KU were in New York, as well. We talked for a while, and Bill and I made plans to get together on the weekend.

Then Bill dropped a bombshell on me. "You know, Bob, that Grace got married last week."

At first, I thought my frat brother was just joking with me. "Oh, I don't think so . . ." I said.

"Hey, Bob! I know so. She married a navy guy."

If Bill had belted me in the stomach with a two-by-four, he couldn't have staggered me any more. I had never suspected that Grace would marry another fellow so soon after breaking up with me. She wasn't even scheduled to graduate until May. My stomach was churning; I felt as though I was going to be sick.

Worse yet, before Bill and I hung up, he also informed me that Bill McCrum, a fraternity brother of ours from Kansas City, had been killed in action, the second Kappa Sig to die since Pearl Harbor. Before the war was over, nearly three hundred young men with whom I had walked the campus of KU would lose their lives. Many others would be wounded, some with lifelong ramifications.

CHAPTER 9

Anticipation

In the military, they say that rank has its privileges. Maybe so, but it just seemed like more responsibility to me. In late January, I was promoted from first sergeant to company commander, which was as high as I could go in the army while at Brooklyn College. The rank carried with it more responsibility, and because it went on my permanent record, I felt that it might be of value to me when I got out of the service. Getting out soon, however, did not seem likely. By February, our original group of four hundred ASTP guys was down to about three hundred twenty, with guys being shipped out to various assignments in the war more frequently now. I had a feeling that my classroom days were numbered.

As part of our physical fitness program, we

had been taking swimming lessons. I'd always wanted to learn how to swim—we didn't have a lot of pools or lakes around Russell when I was a boy—so by learning to swim a little, I felt that at least I had accomplished **something** while in school. By March, I surmounted another major hurtle, getting Bs in most of my classes, an A in something called economic geography, and my only C in mathematics.

More important, I passed my physical exam for combat duty. I was still doubtful that I'd ever make it overseas, but the army had decided that there was a shortage of men in the infantry. It appeared that a number of us who were enrolled in the ASTP were soon to be infantrymen. All that hiking was going to come in handy, after all. Besides being pulled out of the classroom, the drawback was that the infantry required seventeen more weeks of training. But at least we would have guns, which was more than we would have had in the medical corps.

Having completed the first portion of our academic program, the ASTPs at Brooklyn College were given a three-day pass. Realistically, it wasn't enough time to make it back to Russell to see my family, but it was sufficient to catch up on some rest and still do a sightseeing tour of

New York City. Eight guys, five of us from Kansas, did the rounds in New York. We climbed the Empire State Building and had our photograph taken on top of the world's tallest building at that time. From this dizzying height, I glimpsed a sight that will be forever indelibly impressed on my heart and soul—the Statue of Liberty. Something deep inside me said, "That's what this war is all about." Not the statue, but everything that it stood for: the promise of "life, liberty, and the pursuit of happiness" that most of us took for granted on a daily basis.

I needed that extra dose of inspiration a week later, when the army transferred me from Brooklyn College to Camp Polk, Louisiana. A mid-March 1944 letter to my mom and dad hinted at my surprise, as well as my impressions of Camp Polk:

Dear Folks,
From the sidewalks of New York to the swamps of Louisiana; oh, how this Army changes!
They've really kept us busy so far. We arrived here at 3:30 Thursday afternoon, and at 10:30 that night,

we were out in the wide open spaces on maneuvers. They say that we will be here another month, but I hope not.

There are about 3,000 A.S.T.P. men here, and [the drill sergeants] are really taking it out on us. They believe that we have had a soft life for the last three months, so they are either going to make us or break us. I'm in the infantry attached to an anti-tank outfit, but until I learn more about it, I can't explain what it is. . . . This outfit is a little tougher than the medics, but at least we have something to fight back with. They are giving us "Carbine" rifles, and the Anti-Tank Company also is equipped with 57 millimeter guns, which are big and powerful. . . . I'm one A.S.T.P. boy they're not going to break, so don't worry about me. I even sort of like the place.

Good night
    Love,
    Bob

We spent several rainy weeks on maneuvers somewhere outside Camp Polk, and from the first day till the last, it was a constant battle against the mud. We were still living in pup tents when our orders came down that on April 10 our antitank unit would be moving "uptown" to Camp Breckenridge, Kentucky. There, our instruction would focus on anti-tank gunnery operations. I had no idea what that entailed. In telling my folks about Camp Breckenridge, and the latest twist in my military travels, I wrote, "I hear that it's a pretty nice camp. It surely can't be much worse than maneuvers, that's one thing for sure." We were supposed to have six to eight weeks of training in Kentucky, before moving to Camp McCoy, Michigan, where we'd get a chance to fire the really big guns. After that, we might even get to see a little action. Preparing to leave Louisiana, I realized that I'd probably be traveling over Easter. Memories of Easter Sunday in Russell flooded my mind. Bright new dresses for the girls, some of which were made by our mom—sometimes Mom would even wear a bonnet with a flower in it, and of course, Dad, Kenny, and I would put on our Sunday best to go to church Easter morning. Spring flowers, the Empty Tomb, and an emphasis on new life, fresh beginnings, and hope eternal. It all came into focus on Easter Sunday.

The night before Easter 1944, I wrote to Mom and Dad from the swamps of Louisiana, "I couldn't even find an Easter greeting card here. Every time I think of Easter, I think of the fun us kids used to have hunting Easter eggs. I guess Norma Jean will have to find them all this year! Anyway, I'll be thinking of you on Easter Sunday."

That's when I got the good news: between our training at Camp Polk and Camp Breckenridge, we were to be granted a brief furlough. If I paid for it myself, it would be possible for me to travel by train, swing by Russell, and spend a few days at home before going on through St. Louis on the way to Kentucky, as long as I arrived there on time. I was ecstatic. Not only was I going to be out of the mud, but I'd be able to sleep in my own bed and eat Mom's delicious home cooking for a few days.

It poured down rain the entire time I was home, but I didn't care. I walked over to Dawson's to check on Chet and Bub, but of course they were off at war. So was Kenny. Mom and Dad hadn't heard from him for more than a week. They were fairly sure he was in transit someplace. My good friend Bud Smith was still gone, too, but I did get to see Phil Ruppenthal. Most of our other buddies were spread out over Europe and the Pacific. I missed seeing every-body, but it was wonderful just to be able to sit around the dinner table with Mom and Dad. Army life makes every soldier appreciate home.

All too soon it was time to leave for Kentucky. I struggled to keep my emotions in check as Dad and I shook hands on the front porch. I had no idea how long it would be before I'd see him again . . . or what either of us might experience in the meantime.

At Breckenridge, everyone was talking in hushed tones about a possible Allied invasion of Europe coming soon. So, in preparation, we started doing gun drills almost every day. On the range, we spent eight days firing the M-1 ri-fle, the .30-caliber carbine, the .45 pistol, the .50-caliber machine gun, and the bazooka. I was the only one in the entire company who had never fired a gun on a military firing range before. I didn't know what kind of score I'd get, but as it turned out, I did fairly well. We finished up on the range right before Mother's Day. I was hoping that by Mother's Day 1945, Kenny and I would be home to stay—which seemed like a real possibility according to the invasion rumors flying around camp. Some of the guys who had arrived at Breckenridge prior to my group, but after groups such as the 83rd Infantry, which had already shipped out, spoke confidently about the 83rd's destination—someplace in France, a beach they referred to as "Omaha."

On June 1, we started training on the big guns, the .57-calibers. Loud, powerful, cannon-like artillery, the .57s could blow out the side of a building with one shell, stop a tank in its tracks, or provide cover to a platoon of infantry approaching a hostile area. We spent three days practicing with the .57s, and I grew more confident than ever that the war would soon be coming to a close.

It seemed hardly possible that it had been only one year since I'd first reported to Fort Leavenworth for induction into the army. In some ways, the time had passed quickly; in others, it seemed that I'd lived a separate lifetime during the past twelve months. Stranger yet, my brother, Kenny, had been in the army two months less than I, yet he was already on the battlefront. "A year from today," I promised Mom and Dad in a letter, "I'll be home for good—I hope."

Then it happened: D-Day. In one of the bloodiest battles of modern warfare, the Allies landed troops on the beaches of Normandy, France, and fought their way off "Omaha Beach," and into the history books. Code-named "Operation Overlord," the D-day invasion required American and British forces to cross the English Channel for a daring amphibious landing on the French beaches, right in the face of entrenched German firepower. The casualties were enormous—more than nine thousand American and British soldiers were killed, and twenty thousand Germans lost their lives during the invasion—but the attack caught the Nazis by surprise, and within a year, the invading forces took back every square inch of land Hitler's troops had encroached upon.

That night, knowing nothing but morsels of information about the attack, I wrote to my family of my confidence and my concerns:

Dear Folks,

Today was the day everyone has been waiting for—invasion day. It is a little early to predict what will really happen, but the invasion will surely shorten the war. . . . We aren't doing too much here, but they keep us pretty busy with short marching, map reading, and bayonet practice. . . . About all I've done all night is to sit around and listen to the news about the invasion. We are staying up until 10 p.m. tonight to hear President Roosevelt's prayer.

The night before, in one of his "fireside chats," President Roosevelt had informed the American people that "On June fourth, 1944, Rome fell to American and Allied troops. The first of the Axis capitals is now in our hands. One up and two to go!"

Roosevelt went on to give a brief history lesson on Italy's contributions to society, and how Mussolini and the Fascists had enslaved and impoverished the Italian people. He reminded his listeners, "In the north of Italy, the people are still dominated and threatened by the Nazi overlord and their Fascist puppets. . . . Our victory comes at an excellent time, while our Allied forces are poised for another strike at Western Europe—and while the armies of other Nazi soldiers nervously await our assault. . . . It would be unwise to inflate in our minds the military importance of the capture of Rome. We shall have to push through a long period of greater effort and fiercer fighting before we get into Germany itself. . . . Therefore, the victory still lies some distance ahead. That distance will be covered in due time—have no fear of that. But it will be tough and it will be costly, as I have told you many, many times."

I could not have imagined at that point, just how tough or costly it would be.

Then on D-day, June 6, 1944, at ten o'clock sharp, the president informed the world of the Allies' toehold on the western edge of the European continent. Several of my buddies and I gathered around the radio and listened intently.

Speaking from the White House, Roosevelt said, "Last night when I spoke with you about the fall of Rome, I knew at that moment that troops of the United States and our Allies were crossing the Channel in another and greater operation. It has come to pass with success thus far. And so, in this poignant hour, I ask you to join with me in prayer."

And then Roosevelt did what few American presidents have dared to do since FDR's time. The president of the United States **prayed**—in public—out loud—on nationally broadcast, public airwaves. "Almighty God: our sons, pride of our Nation, this day have set upon a mighty endeavor, a struggle to preserve our Republic, our religion, and our civilization, and to set free a suffering humanity.

"Lead them straight and true; give strength to their arms, stoutness to their hearts, steadfastness in their faith.

"They will need Thy blessings. Their road will be long and hard. For the enemy is strong. He may hurl back our forces. Success may not come with rushing speed, but we shall return again and again; and we know that by Thy grace, and by the righteousness of our cause, our sons will triumph.

"They will be sore tried, by night and by day, without rest—until the victory is won. The darkness will be rent by noise and flame. Men's souls will be shaken with the violences of war.

"For these men are lately drawn from the ways of peace. They fight not for the lust of conquest. They fight to liberate. They fight to let justice arise, and tolerance and goodwill among all Thy people. They yearn but for the end of battle, for their return to the haven of home.

"Some will never return. Embrace these, Father, and receive them, Thy heroic servants, into Thy kingdom.

"And for us at home—fathers, mothers, children, wives, sisters, and brothers of brave men overseas—whose thoughts and prayers are ever with them—help us, Almighty God, to rededicate ourselves in renewed faith in Thee in this hour of great sacrifice.

"Many people have urged that I call the Nation into a single day of special prayer. But because the road is long and the desire is great, I ask that our people devote themselves in a continuance of prayer. As we rise to each new day, and again when each day is spent, let words of prayer be on our lips, invoking Thy help to our efforts.

"Give us strength, too—strength in our daily tasks, to redouble the contributions we make in the physical and the material support of our armed forces.

"And let our hearts be stout, to wait out the long travail, to bear sorrows that may come, to impart our courage unto our sons wherever they may be.

"And, O Lord, give us faith. Give us faith in Thee; faith in our sons; faith in each other; faith in our united crusade. Let not the keenness of our spirit ever be dulled. Let not the impacts of temporary events, of temporal matters of but fleeting moment—let not these deter us in our unconquerable purpose.

"With Thy blessing, we shall prevail over the unholy forces of our enemy. Help us to conquer the apostles of greed and racial arrogancies. Lead us to the saving of our country, and with our sister nations into a world unity that will spell a sure peace—a peace invulnerable to the schemings of unworthy men. And a peace that will let all men live in freedom, reaping the just rewards of their honest toil. Thy will be done, Almighty God. Amen."

Tears clouded the eyes of several of the soldiers sitting near me, listening to Roosevelt's prayer. A few sobbed openly. Historians would one day debate Roosevelt's policies, his strengths and weaknesses, but that night, in that moment, his voice transcended human expressions. Somehow, he had tapped into the soul of America, and had expressed its cry without reservations or embarrassment, to God.

Every soldier in uniform who heard Roosevelt's words and was moved by their meaning had no doubt that we were going to win the war, no matter what the cost.

CHAPTER 10

You Just Had to Be There

The war seemed to shift into high gear after the D-day invasion. Everything changed as Allied troops continued the successful, though agonizingly painful assault on France; yet ironically, for those of us still in training, preparing to get into the battle, life remained oddly routine. The Sunday after D-day I got up and went to church, then went off base to a "picture show" at the local theater, and spent most of the hot afternoon in an air-conditioned drugstore and the USO.

When I received a letter from the folks back home informing me that another hometown boy, Ralph Tichenor, had died, reality and the horror of the war ripped at my heart. I hurt for the Tichenor family, yet something told me deep inside that Ralph wouldn't be the last; there would be more Russell casualties before the war ended.

Still at Camp Breckenridge learning about artillery, by mid-June, I started to get bored and anxious again. I'd heard that there may be some openings in OCS—I'm not sure that it completely dawned on me just **why** there'd be openings at that point in the war—but I began thinking again about applying to become an officer.

Besides our regular physical conditioning, I was training to compete in our regiment's track meet coming up in early July. Each evening, sometime between dinner and lights out, I'd put on my gym shorts and track shoes and go out running. For those unfamiliar with the military, it may seem strange that the army would be conducting track meets during wartime, but the extracurricular sports activities were a standard part of morale building and tension release, even for troops overseas relatively close to the front.

One day, I had just finished working out and I was dead tired. I ran down past the barracks, where a friend with a camera was taking some pictures. I was all sweaty, and wearing only my track shoes, socks, and shorts, but my buddy stopped me and took a picture of me in front of one of the big .57-millimeter guns, the kind designed to knock out tanks. I looked tough and strong, in tip top physical condition; but had I known how many people would eventually see that picture, I might have cleaned up a bit first.

On June 18, I called home to talk with my folks about OCS. I dialed the operator at three o'clock in the afternoon, and at nine o'clock, I was still waiting to be connected. Finally, I gave up. Nowadays we take for granted the simple act of picking up the telephone and calling someone across the nation, or around the world. In 1944 it wasn't so simple. A call from Breckenridge to California could take as long as eleven to twelve hours before being connected.

I had wanted to tell Mom and Dad about deciding to take another shot at getting into Officer Candidate School. I wrote to them instead:

I am supposed to be interviewed tomorrow, but I haven't much chance because they are only taking four men from our regiment, and there are at least 75 applications. I applied for infantry O.C.S., and I'm afraid that I haven't had enough infantry basic. I

**put down Medical Administration for
my second choice.**

In the meantime, I was learning more about
the various weapons available to the infantry.
For a full week, I practiced firing a .50-caliber
machine gun on the Breckenridge range. I had a
lot of fun firing the machine guns, even though
my ears rang for hours after each practice session. I'd never dreamed how powerful those
guns were. I also practiced firing the carbine,
the M-1 rifle, and the .45-caliber pistol, and I
learned how to throw hand grenades. There
really is an art to dealing with hand grenades, I
discovered. The only part of the gun training
that I didn't care for was cleaning them, a discipline that the army strictly enforced.

In my spare time I tried my best to keep up
with the mail, and even teased Mom for not
writing more frequently. "Are you out of stamps
or something?" I wrote to her in late June. "I
haven't heard from either you or Dad for over a
week."

I scored seven points in the big regimental
track meet in July, placing second in the one-hun-
dred-yard dash, second in the 220, and I ran a leg
of a four-man relay, too. I earned a three-day pass

for my efforts, although the way things were looking, I wasn't sure I'd get a chance to use it. I had been informed that I might be getting shipped out to Fort Benning, Georgia, to the Benning School for Boys, as it was derisively referred to by the old-time officers. At long last, I had been selected for Officer Candidate School.

Writing home, I informed Mom and Dad that I was about to have another new address:

I hope that I'll get to leave tomorrow for Fort Benning, Georgia. I haven't been told for sure, so I'm keeping my fingers crossed. I had my physical Wednesday and my interview Thursday, and passed both. They told me not to leave the company area Saturday or Sunday, for I would probably go then.

I'm going to infantry O.C.S. so it will be plenty tough, since I haven't been in the infantry long enough to know what it's all about.

A few days later, my hopes became a reality. On July 18, 1944, I wrote the family on plain beige paper from my new home:

Dear Folks,

Here I am in Fort Benning, Georgia. We arrived last night and have been going through processing since then.

I've taken two tests, filled out three or more papers; I take a physical and have my teeth checked in the morning.

I was made a corporal when I left Breckenridge, since you have to be a non-com to go to O.C.S. There is everybody from First Sergeants on down going to O.C.S. here. We are all in the same boat, though. A First Sergeant pulls K.P. just like everyone else.

. . . It's really a tough school, both physically and mentally, and since I haven't had much infantry training, don't be disappointed if I wash out. I'll try my best and keep my fingers crossed.

Washing out was a distinct possibility. By the time we completed another seventeen weeks of training, nearly half of the soldiers who had qualified for OCS would be gone. Nevertheless, I was greatly encouraged after the third week, and I was still hanging in there. I had even scored the highest grade in our company on a physical achievement test. I needed it, too. My grade on the machine gun test was lousy.

By mid-August the war news was sounding good. The Russians were conducting frequent bombing raids in Germany and making big advances on the ground toward Berlin; the end couldn't be far away now. I told someone that if things kept up that well, we'd be home in a month or two.

Fat chance. Instead, we finished our course on using the bayonet and the Browning automatic rifle. I tested well on both of those, and on another physical achievement test I remained among the top ten men in the company. I had lost nine pounds since arriving at Fort Benning, and was down to about 182 pounds. I figured it was due to the intensely hot, humid Georgia summer, and that I'd put the weight back on at the first sign of a chill.

The weapon I had the hardest time mastering was the mortar. I was having trouble catching on to several other weapons used by the infantry, but for some reason the mortars were a special pain to me. In early September, I felt sure that I had failed a test on mortars, and I was disgusted with myself. Imagine my surprise when I received my paper back and saw that I had scored a B on the test. I almost fainted. I guess I was lucky on that one.

I wrote home and suggested to Dad that he pick up some shells so we could go hunting when I got home from the army. Now that I had trained on everything from an M-1 rifle to a grenade launcher, I felt confident that I'd be a better marksman than I'd previously been in civilian life. In my letter, I told the folks, "I don't need a thing, except some good homemade fudge."

Around the eleventh or twelfth week, our training intensified. So much so that I didn't have time to write home for more than three weeks. We had several six-day bivouacs, as well as two "night problems," simulated night patrols—going out looking for enemy movements in the dark—that began at three o'clock in the morning. I was exhausted. When I finally wrote to Mom and Dad, I started with an apology:

Dear Folks,
   I'm almost ashamed to write home because it's been about three and a half weeks since I've written, and I

imagine that you are pretty well disgusted with me.

A lot has happened the past month. We had our 12th Week [qualifying] Boards, but I didn't have to go before it. We lost 13 men today. Monday we have our 14th Week Boards, and will probably lose 10 or 20 more. We have 127 men now out of our assigned 200, so I'm pretty lucky to still be here, I guess.

We had our last physical achievement test three weeks ago, and I received a superior rating and was first in the company. . . .

Received the package Monday, Mom, and everything was good, as usual. The picture of you is really nice. Now I have a girl to set on my bookshelf. I haven't had one since Grace was married. Tell Dad not to forget to have his taken, too. . . . I just heard that KU beat Nebraska, the first time since 1917, pretty good, eh.

Good Night,
    Love,
    Bob

The months of October and November were a blur of activity. The Army seemed obsessed with getting my class of OCS guys through the program and into the action. We trained night and day, strenuous days packed with marches and combat practices, classroom lectures on the use of mortars and machine guns, field instruction on everything from the most effective way to hurl a hand grenade, to how to operate an anti-aircraft gun.

By this point in the war, the Allies' inch-by-inch progress on the Pacific front, and the invasion of Europe, had taken a frightful toll on young American soldiers. One group that was diminishing faster than it could be replenished was the pool of second lieutenants. Fort Benning was cranking out new officers as fast as they could train them, without lowering the standards of preparation. The program seemed almost too fast, too compressed. I arrived in August as a corporal, and in little more than ninety days, I left in November as a lieutenant. Whether I merited his confidence or not, the commanding officer at Fort Benning believed in me, and declared that I was competent to handle a platoon of men, so I didn't argue with him. I believed.

I felt a nervous excitement as the first week of December began. After training for more than a year and a half, it appeared certain that I was finally going to where the action was. That was exciting. On the other hand, if I did get into what was left of the war, it was sure to be intense. This would no longer be target practice; these targets would be shooting back in life-or-death, last-ditch frustration and desperation.

I received a new APO, Army Post Office address, in care of the Postmaster, New York City. This was a clear sign that my days at Fort Benning were numbered, that I'd be shipping out soon. "Send any personal belongings or extra clothing home or lose it," the commander ordered. "Be ready to move out at a moment's notice," were our instructions.

The notice came on December 4, 1944, so I dashed off a quick note to Mom and Dad.

Haven't much time before leaving, and I wanted to send you the key to my footlocker and the paper on my Class "E" allotment [sort of an Army bonus payment]. You and Dad can do anything with the money that you

**like, for I owe you much more than
the allotment will ever pay. . . .**

   **Got to go now; we leave here in
sixteen minutes. Tell everyone "hello"
and "good-bye" for me.**
   Love,
   Bob

Along with a hundred or more other young officers, we traveled by train from Georgia to our port of embarkation, in Fort Meade, Maryland. When we arrived at the dock area, a band was playing John Philip Sousa marches, onlookers were waving small American flags, and the Red Cross ladies were passing out free coffee. The port area bustled with activity: soldiers, sailors, and airmen everywhere, some parents and other well-wishers waving from behind the lines, young women with tears in their eyes, young men stoically hugging and kissing the girls good-bye one last time. Other soldiers were walking up the gangplank to board the huge troop transport ship that would take them to the war they'd only imagined in their best dreams or worst nightmares. It was a day we had been waiting for and dreading at the same time, loving and loathing, anticipating it yet hoping it would never come.

The men were boarding by groups. When our turn came, I slung my duffel bag over my shoulder, took one last look back at the dock, and walked up the gangplank with my group. We stopped aboard the first deck. "Dole!" I heard the processing officer call my name.

"Robert J.," I responded, as I took my place on board the ship.

The ship's quarters were extremely cramped, and it was obvious that everyone was a little nervous. But the tension was soon broken by the usual banter of soldiers joking and telling stories. I'd never been on a ship before, so I was fascinated by the entire experience. As I watched the tons of equipment loaded onto the ship along with the hundreds of soldiers, I looked down at the cold, choppy waters and wondered how much more we could take on and still stay afloat.

I needn't have worried; the ship was massive. Soon it was time to depart, so I climbed the stairs and stood on the highest deck I could get to and, along with hundreds of other soldiers, waved at anyone on the dock who waved back as the giant vessel began to move. With the help of several tug boats, the enormous troopship pulled away from port and eased on out into the Atlantic Ocean with far less jostling than Mom's driving. The waters turned a darker shade of blue, then almost black as the transport ship headed across the Atlantic toward the Mediterranean Sea.

Our orders said "Italy," although I didn't know much about where I'd be stationed or what conditions I might be facing. I understood, however, that I'd be relieving officers who had been on the front for a while, so I expected to be thrown into the fracas in a hurry.

The trip across the ocean would take approximately two to three weeks, so I passed part of the time by playing cards with the guys, and part of the time writing letters. Naturally, one of my first letters was to my parents. For security reasons, I was not permitted to tell them where I was going or from what port I had embarked. But I gave them a hint as to where I was headed by referring to Eugene Ruff, a young man who lived across the street from us in Russell and was now an officer someplace in Italy, not far from Rome.

Dear Folks,

**I told you in my last letter that I
would try to write again before
leaving. We are aboard ship now; can't
say where or when we boarded. It
looks as though I won't get to see
Kenny now, but I may get to see
Eugene Ruff in the future. . . . Don't
worry about me though. I'm feeling
fine, and hope everyone at home is
feeling as well. . . .**

Good-bye folks,
Love,
Bob

A week or so into the trip, the Atlantic waters
turned vicious. Huge waves rolled across the
ocean, and as large as our ship was, it still
heaved up and down and rocked from side to
side. My stomach did, too. I had heard the saying, "Kansans don't have sea legs," and now I
was experiencing the truth of that statement.
Putting it discreetly, let's just say that I gave a
whole new meaning to the term "army green."

Finally, as we neared Europe, the seas calmed
and I was able to keep some food down, as well

as write a few more letters to the folks at home. I didn't really have any new information, and I had told as much as I dared put in a letter about the ship and the men on board, but I didn't want to miss the chance to send a note home. On December 19, I scrawled a quick letter:

Dear Mom, Dad, Norma Jean,

I wrote last night and nothing has happened since, but I know that you feel better when you hear from your sons in the service, so I'll write as often as possible.

Every day we get a little closer to our destination, and I'll be glad to get my feet on the ground again, even though I do sort of like the ship I'm on.

It was a weird feeling, watching the waves lapping against the ship and feeling the wind pick up as we steered toward the Mediterranean coastline, wondering if we'd get there in time even to fire a shot, hoping that the Germans would surrender soon.

A few days before Christmas 1944, our ship carefully eased into war-ravaged Naples harbor, where I saw for the first time the grim physical realities of war—sunken ships in the outlying waters, piles of rubble where beautiful buildings had once stood, and perhaps worst of all, young children wearing clothes that looked like rags, as they waved at the United States soldiers and begged for food. Mount Vesuvius, the volcano that had buried the nearby city of Pompeii in A.D. 79, loomed in the background of the city.

We disembarked into landing crafts that took us to shore. By now I was glad to get off the ship and on solid ground again. I stared with unabashed wide-eyed innocence at the noisy, frenetic sights and sounds of Naples. An old city filled with religious construction of baroque and rococo-style architecture, Naples had been one of Hitler and Mussolini's early objects of desire, due to its large harbor, second only to Genoa's, in Italy. The city had paid a high price for such prestige, being incessantly bombed and battled over, with whole areas turned into drab beige and gray rubble. Fortunately, some of the oldest landmarks and churches survived, leaving a testimony for future generations by their mere existence.

Nevertheless, seeing the horrendous devastation of war close up for the first time left scars that will never be healed. It's difficult to describe the utter destruction of war, what it does to the physical surroundings, the land structures, buildings, highways, bridges—turning beautiful homes, stores, schools, offices, and churches into dirty gray nothingness—piles of annihilated matter that formerly had something to do with people's daily lives. It's even tougher to describe what war does to a human being. Mere words seem so inadequate a vehicle to convey that gut-wrenching feeling a person experiences looking at the realities of war. Nothing does it justice, and that's why many soldiers don't even try. With a wave of the hand and a glassy stare into space, they'll say something like, "You just had to be there." Not that they can't describe it. They simply don't want to go back there to the killing and death, the sights, sounds, and smells of war, the bombed-out towns, doorways that open into nothing but rubble, images of shredded bodies, the vestiges of which will live forever in their hearts and minds.

We hiked to our staging area outside Leghorn (Livorno), then the troops that had come over together on the ship were disseminated to various locations in Italy. The winter weather was still cold and wet, with plenty of snow on the mountains and plenty of rain in the valleys, making movement slow, muddy, and miserable. But as the weather cleared and we were able to unpack our equipment, I became more acclimated to the camp. Before long, I was sent on to the 24th Replacement Depot, far out in the countryside, about halfway between Rome and Florence.

I checked in, was processed, and was assigned a tent, some straw, and some blankets to shield me from the cold. Like the other young officers, I was warmly dressed, and we had plenty of food at the replacement depot, so for the time being it seemed like the optimal place to be. We had lots of time to ourselves, and life at war didn't seem so bad.

Settling in, we even celebrated Christmas at the depot, and for a few hours, the war seemed far, far away.

I wrote home shortly after New Year's Day. I wanted to let the family know that I was okay, but also that my new circumstances weren't going to get me down. Looking back, it's easy to imagine that I was smiling as I wrote this:

Dear Folks,
I haven't written for three days, so here goes my first letter of 1945. Hope

you all had a Happy New Year. We had turkey for dinner; there wasn't much, but it was good.

I've visited Florence since I arrived. It really is pretty nice since it hasn't been bombed much. . . . You can send me something to eat whenever you're ready. Send candy, gum, cookies, cheese, grape jelly, popcorn, nuts, peanut clusters, Vick's vapor rub, wool socks, wool scarf, fudge cookies, ice cream, liver and onions, chicken, banana cake, milk, fruit-cocktail, swiss steak, crackers, more candy, Life Savers, peanuts, piano, radio, living room suite, record player, and Frank Sinatra. I guess you might as well send the whole house if you can get it in a five-pound box. I would like some food though, honestly.

The only drawback to my assignment that I could see was that there was so little to do at the replacement depot, we nearly bored ourselves to death. We spent our days studying maps, cleaning weapons, and trying to appear busy. Truth was: we were playing a waiting game—waiting for the war to end on one hand, while on the other hand, we were sitting around waiting to be replacements . . . somewhere.

Meanwhile, the German propaganda machine continued to churn out messages aimed not so much at extolling the glories of the Third Reich as attempting to demoralize young American soldiers who were far away from home and their loved ones. The Germans air-dropped thousands of leaflets. The leaflets read, "Today you are stuck in the mud, and must fight thousands of miles away from your country in the rugged mountains of Italy, for those who are still enjoying going places, and doing things." A fake **Life** magazine cover depicted a seductive young woman with the ad copy, "Who knows whether you'll ever make whoopee again!"

An even more gut-wrenching aspect of life at the Twenty-fourth Replacement Depot was the inability of the officers to form any lasting friendships. Almost daily, another second lieutenant went off to take over a platoon somewhere, to replace an officer who needed a rest or, worse yet, had been killed or wounded in battle.

We were able to get good radio reception at the replacement depot, and the best part about this was that we could keep up on the war news. I probably surprised Mom and Dad when I wrote home to them with more up-to-date information than they had:

Just listened to a news broadcast and it sounds as if the Russians are really going places. I hope they get there in a hurry so this war will end. Have you heard from Kenny since the Americans started their big drive in the Philippines? I suppose he is still in New Guinea, but I've been wondering, ever since the invasion of Luzon.

I received very little mail from home during my first few weeks in Italy. Not that the folks weren't writing; it just took longer for their letters to get to me. When the mail finally arrived, it was a mixed blessing. The good news was that Kenny was safe and well in the Pacific, and had been promoted to the rank of sergeant. I was really proud of him, despite the fact that he owed me a bunch of letters. I quipped in a note to Mom and Dad, "He'll probably be a General by the time I hear from him. I'm probably as high as I'll go for three or four months, but I'll be satisfied if my next promotion is from Second Lieutenant to a civilian."

I was excited to receive any news from home—few things are more disheartening for a soldier than to go to mail call every day and never get a letter—but not all the news from home was good. Mom and Dad informed me that two more Russell boys, Vernon Landon and Reuben Heffel, had been killed in combat. I knew that the death of two young men at war was not shocking news to the rest of the world, but in a small town like Russell, the loss of those two young men sure made an impact.

In late February 1945, my name was called. The army assigned me to the front in the Apennine Mountains, to take over a platoon, replacing an officer who had been with the 10th Mountain Division. My commanding officer didn't tell me much about my assignment, just that the 10th Mountain Division had recently been involved in some intense battles in their successful efforts to take the strategic location of Mount Belvedere, south of Florence. I'd be stationed outside the quaint little village of Castel d'Aiano, a town of about four thousand people that had been occupied by the Germans for much of the war. Had it not been so severely battered by bombs, the mountain town would have sounded almost idyllic, like a name you might see on a tourist brochure for a lovely vacation spot.

But this was no vacation, and the Nazis entrenched in the bunkers on every rise in the area were no welcoming committee.

CHAPTER 11

Castel d'Aiano

If you've ever seen the movie **The Sound of Music**, you have a pretty good idea of what the region surrounding Castel d'Aiano looked like in 1945, and for the most part, still looks like today. Picturesque, with snowcapped mountains and lush green valleys, divided by spring-fed cold-water brooks, interrupted only occasionally by the still-working ancient water mills, the rugged terrain posed numerous farming challenges to the local population. To survive, they grew potatoes, mushrooms, white truffles, legumes, and other hearty vegetable crops; most families owned their own cows for milk and cheese, and many homes had a chicken coop just outside the back door. Thickets of chestnut trees reluctantly yielded their hard-shelled nuts for snacks. Dirt and gravel-bedded roads snaked back and forth up the mountainside, more easily traveled by mule than by tanks, jeeps, trucks, or other military vehicles. Stone farmhouses, some of which dated back to the 1200s, dotted the hillsides; others were built into the cliffs in ways that sometimes defied the forces of gravity.

Indeed, the people of Castel d'Aiano were accustomed to defying powerful forces. Much of the town's social and spiritual life continued even during the Nazi occupation, centering on the Catholic church, with its sanctuary that seated about two hundred people and its centuries-old bell tower, which continued to call people to worship until the day it was obliterated by Allied bombs.

Standing on the mountaintop and gazing at the breathtaking, panoramic beauty of Castel d'Aiano's natural environment, as I've done several times since the war, you find it hard to imagine why any army would dare tamper with the peace and quietness of that place. But when you look at a map, you understand why the Germans coveted this high ground and fought for it so tenaciously. Just southwest of Bologna, Castel d'Aiano is located in the heart of what the Allies referred to as the Gothic Line, the long line of German defenses dug in on top of the mountains, extending across the top of northern Italy from the Adriatic Sea on the east, all the way to the Mediterranean Sea on the west of the country. The Nazis referred to the Gothic Line as the Green Line, since their artillery bunkers were tucked into the tree-lined mountaintops. Actually, the Gothic Line was not one line of defense but two, along a chain of the Apennines, and backed up by another long chain of the Apennines known as the Genghis Khan Line, both making full use of the natural high terrain to create an almost impenetrable barrier between Italy's "boot" and the rest of Europe.

But once Rome fell in June of 1944, the Allies continued to push northward, punching bloody holes in the Gothic Line, and by February 1945 the Germans were reeling, retreating from one mountaintop to the next. Not far from Castel d'Aiano rose Mount Belvedere, one of the highest peaks in the region, and one of the strategic Nazi defenses, blocking access to the Po River Valley, beyond to Bologna, and a straight shot into Hitler's heart.

The Nazis fully expected the Allies to attack through the valley, and they were ready. But the Germans never imagined the cunning, skill, and tenacity of General Hays and the 10th

Mountain Division. Thinking the sheer cliffs of Riva Ridge were inaccessible, the Germans were caught by surprise when the 10th Mountain Division scaled the ridge during the night, helped by a lingering fog in the early-morning hours. Once Riva Ridge belonged to the Allies, Mount Belvedere, though a formidable challenge, became vulnerable, and by late February, the Allies owned the highest peak in the area.

On March 4, 1945, the town of Castel d'Aiano was liberated by the Americans. The Nazis retreated to the steep hills outside the town. They had taken over the area early in the war, literally moving in with the Italians—confiscating their homes, possessions, food, and women. Now, they hid wherever they could, threatening to slaughter the Italians if they revealed the Nazis' locations. By the time I had arrived in February to take over the Second Platoon of I Company, nearly every home in the hills north of town had Nazis living in the "cantina" (the cellar) and machine gun emplacements in the bedrooms and attics. The American forces, along with a regiment of Brazilian soldiers and the patriots (local Italian freedom fighters), were camped on one side of the mountains, in an area known as Monte della Spe. The Nazis were dug in on top of a steep hill directly opposite us, less than a mile away, a hill hosting a few remaining stone farmhouses that had somehow survived the bombs—a hill the Allied mapmakers had designated simply as Hill 913.

The next few weeks were extremely busy as I immersed myself in my new responsibilities. On March 13, I found enough time to send a note to Mom and Dad.

Dear Folks,
 I hope you haven't worried too much because you haven't received any mail from me. I think it's been about two weeks since I've written. I've really been keeping pretty busy. I've got a platoon of men to look out for, and it seems that I never have much time to myself. When I'm not busy, I've generally got a big stack of mail to censor.
 I'm a combat soldier now folks. I suppose you've been reading about the 10th Mountain Division in the paper the last few weeks. We've really done some pretty good work so far, and I hope we can continue to do as

well. I'm feeling better than ever so far. I guess this outdoor life agrees with me. I'm starting to get a sun tan since the weather has warmed up.

We are in a rest area now, so whenever I do have a chance to write, you'll know that I'm writing from a rest area. It is really nice around here. We are living in a hotel, and I really enjoy sleeping nights in a bed instead of a foxhole. A foxhole isn't as bad as you probably think. We generally fill the bottom with straw which makes it pretty comfortable.

I'm sorry to hear about all the Russell boys being killed or wounded, but I'm glad that you write and tell me anyway. I guess so many were meant to be killed in this war; there's nothing either you or I can do but trust in God, and I pray that he will look after us.

A few days later, on March 18, I was on a night patrol when we drew fire from some Nazis moving around the chestnut thickets in the dark. That's when the grenade fragment cut into my leg. I'll never know whether that grenade was thrown by me or someone else in our platoon, and it really doesn't matter. Several other guys and I were injured by the blast, but fortunately, all of us were able to be patched up and put back together. The army even awarded us Purple Hearts, for having been wounded in action. A lot of guys who ran into Nazi ambushes in the dark weren't so lucky.

The German artillery emplacements, dug into pillbox bunkers in the hills, were shelling us every day now. It was almost as though they could sense that the end was near, but they were not going to surrender. We didn't know until much later that Hitler had ordered them to defend the Po Valley to the last drop of their blood.

When I wrote home on March 21, I glossed over the injury that I'd suffered a few days previously. I told Mom and Dad, "When I'm not in my foxhole ducking German artillery, I'm generally on night patrol or trying to catch some sleep. I'm pretty dirty right now. I haven't showered or washed for several days, but I guess it won't hurt me any. I can't see why the Jerrys won't give up, but I'm just a second lieutenant who doesn't know too much about the big picture."

Mom responded immediately when she learned that I'd been injured. I knew she was concerned, but she expressed it with her characteristic can-do attitude and tongue-in-cheek humor, almost glibly referring to my eighteen months of "classroom training," only to be wounded on one of my first nights out on patrol. She wrote:

> **Hi, Son,**
> **What's this I hear about patrols and grenades, etc. Boy, we were sure sorry to hear of the mishap. . . . I suppose you're back in a rest camp now. Where did they get you? There must not have been a school solution to that situation.**
> **They didn't waste any time getting you in the scrap, did they? Well, I hope you won't have to go back too soon, and I hope the injury wasn't serious. . . . By the way, what do you plan to do when this war is over? Go back to school or what? . . . Don't know if I told you in my last letter that I planted a Victory Garden. Anyway, it's really growing. With**

Dad's advice and my sweat, we'll make a go of it. Course, there's a reason for planting it. You see, I'm expecting a visit from you when you get back, and I plan to do a lot of canning. Course, I'm not tight or anything. But that way, I won't have to charge you for the meals, and I won't have to worry about my grocery bills quite so much, and that way, too, it will be a much more pleasant visit for both of us! Ha!

Does that sound like a mother, or what? Her message between the lines was: you may have been scraped by a grenade fragment, but I'm expecting you home soon, so don't be taking any unnecessary chances.

Of course, she didn't really need to remind me of that.

It was a welcome relief when our unit got a break and we were able to move back to a rest area for a few days. About that time, some guys in the air corps offered to fly me from Florence to Rome during our down time. I had never before flown in a plane, so I wasn't going to turn down that chance.

Since Rome had fallen to the Allies in June 1944, just before D-day, the city was now swarming with American soldiers. Still, I felt like a real tourist in Rome; I enjoyed doing a little sightseeing, getting around to many of the ancient landmarks such as the Roman Forum and the Colosseum, as well as the Vatican, and I was thoroughly impressed by the city's beauty and architecture.

One of the most popular places among American soldiers was the Excelsior Hotel on Rome's famous Via Veneto. The extravagantly gorgeous hotel had been the Nazis' headquarters prior to the city's liberation. Replete with dark wood-paneled walls and high ceilings from which hung enormous crystal chandeliers, the elegant hotel also housed a coffee shop, where I purchased a warm drink and then walked across the street to sit down in an outdoor café and "people-watch." It was the best show in Rome, as the potpourri of strolling musicians, bustling businessmen, and pretty girls passed by. My buddies and I nursed our drinks for hours as we enjoyed the sights.

Best of all, while in Rome, I ran into two friends from the University of Kansas, Dean Nesmith and Bill Hargiss. Dean had been an athletic trainer at KU for all sports, including Phog Allen's basketball teams. A tough guy and a no-nonsense sort of athlete, Dean was serving with the army's Sparta School, a special behind-the-lines sports program for guys in the war zone.

We talked a little army shop, but mostly we talked about the latest news from KU, and especially the Kansas Jayhawks' recent football victory over the Nebraska Cornhuskers. Just talking to the guys from KU whet my appetite to return to school, and more specifically, to return to the competition of Phog Allen's basketball team, not to mention football and track. Of course, there's nothing like living in a cold, filthy foxhole for a few weeks to make a guy long for the pristine environment of a university classroom. And what better landscape to remind me of the well-manicured lawns of KU than the bombed-out, pockmarked craters carved out of the hills and valleys around Castel d'Aiano?

When I told Dean where I was stationed, he was intrigued. He knew how much I loved sports, and the wheels in his head must have started turning quickly since I was already in the area. "Man, we need to get you over here in Rome, with us," he said. "It's only a two-week program, but you'd love it. Who knows? If you spent a couple of weeks in the sports program, by that time, this war might be over, and we could all go back to KU."

Dean offered to check on whether he could get me into the program. I readily agreed.

"I'm just on a short leave," I said. "I have to head back to Florence soon, and then back out to my foxhole." Dean said he'd do his best to see what we could work out for the upcoming class session, and would let me know.

I was more enthused about going back to school than I had ever been. On April 4, when I wrote to Mom and Dad, my excitement was obvious:

Well, I have plenty of time to write for a few days, unless I do too much sight-seeing. I finally got a break, and wound up in Rome. It's really swell here, almost like home, but it won't last long.

I was very fortunate today. I ran into Dean Nesmith and Bill Hargiss, who are in Rome conducting a sports program. Nesmith was our trainer at KU, and Hargiss was my track coach.

I've spent the entire day visiting with
them. I've a pretty good chance at
going to the next class. If I do, I'll be
back in Rome before too long. Maybe
you've read about the Sparta School
in the K.C. paper. A class lasts two
weeks and is composed of sixty men,
half officers, half enlisted men.
During the two week period, you are
taught practically every sport in the
business. What they are trying to do is
develop instructors for the post war
world. If I'm lucky enough to get in, I
will let you know. I'd sure like to relax
for two weeks, with nothing to do but
play basketball, football, and run. . . .

When I saw Coach Hargiss and my
old trainer Nesmith, I was convinced
that I'm going back to school.
Primarily to study, but also to
compete in athletics. I love to play
basketball and football, and compete
in track meets. I hope I'm not too old
when the war ends, for I really want
to do a lot of things.

I weighed myself today. I tip the
scales at 193 pounds, which is more

**than I've ever weighed before. I guess
the "C" rations must agree with me.**

Mom must have received my letter somewhere around the time President Roosevelt
died. The president's death was a heavy blow to
her, and like the troops in the trenches, she no
doubt shed many tears over our fallen commander in chief. On April 14, 1945, my mother
wrote to me:

**Dear Bob,
   I was surely glad to hear you had a
leave, and very happy to know that
you ran into some of your friends you
know so well. And I do sincerely hope
with all my heart that you get the two
weeks of school you spoke of, because
I know how much you love that kind
of sports. And while you're relaxing, I
can relax, too. So we can all have a
little peace and rest, and by that time,
we hope and pray this will be over.
They look for it hourly over here now.
   I hope you thoroughly enjoyed your
stay in Rome and I'm glad you had
a safe journey from Florence to**

Rome. . . . I'm especially glad you're thinking of going back to school. . . . I bought five pounds of homemade bittersweet peanut cluster, and I'll mail it Monday. It is really delicious. I sure hope you get it in a hurry, and I do hope you get some of your packages soon. There's a lot of them on the way, son.

The Nation is certainly mourning the life of a Great Man. I hope the morale of you boys overseas isn't broken by this news, and I hope Germany and Japan don't feel that this gives them any more hope of winning this war. I sincerely hope that Truman carries on in the same manner, but I'm sure we've lost a president that our future generations are going to know as we knew, and learned of Lincoln and presidents of his equal. I'm afraid as it so often happens, we didn't really, any of us, realize how great a man he really was.

The war news sounds like it is about over. Russia is 13 miles from Berlin; that was noon today. Bob, do be

careful. I don't really believe this
thing will last very much longer. I
know you'll get home soon—so keep
your chin up. . . .

Good luck, son. Bless you. May your
wishes all be fulfilled. I'm thinking
and praying for you always.

Love,
Mom

Little did my mother know that as she penned
those words, I was lying flat on my back on the
battlefield near Castel d'Aiano, along the side of
the mountain, on Hill 913, in the same spot
where Sergeant Carafa and Stan Kuschik had
left me six hours earlier in the day. I was fading
in and out of consciousness, almost delirious
and not far from death's door. I was going to
need a lot of Mom's prayers.

CHAPTER 12

Just Give It Time

"Lieutenant. Lieutenant! Wake up, we've got to get you out of here."

When Dorothy awoke from her dream, she was back in Kansas surrounded by her loved ones. But this was no dream, and I was still lying on the ground, faceup, with occasional raindrops spattering off my face. My arms remained crossed over my chest, just as Sergeant Carafa had left me six or seven hours earlier. I couldn't move. I couldn't feel my legs or my arms. The afternoon light was beginning to fade as evening set in, and I had difficulty telling the difference between the fading daylight and my own intermittent consciousness. That's when I heard the voices of men standing over me. I was glad they were speaking English, not German.

Apparently, in his wounded, woozy condition,

Arthur McBryar had wandered around until he had found help and pointed out to the medics where I was lying. Thankfully, the Americans had spotted him before the Germans could pick him off. I learned much later that two other medics who had set out to help us earlier in the day had themselves been hit by enemy fire.

The 10th Mountain Division's 85th Regiment battled all day long and part of the next to reach the top of Hill 913 and the hills adjacent to it, opening the way to the Po Valley and beyond. So many guys from I Company were killed or wounded, that few of the soldiers in my platoon ever got to see the panoramic vista from the summit. Instead, K Company came alongside and finally ascended the heights. Dislodging the Germans from the high ground came at a horrendous price. The Americans suffered more than 460 casualties on April 14, with 98 men dead following the daylong assault on Hill 913. Any German soldiers who were not killed or seriously wounded ran for other hills, with the American attackers in hot pursuit. The Germans were afraid to surrender; after all, Hitler had ordered that they fight to the death. Moreover, they had heard that the 10th Mountain Division had a reputation for taking no prisoners, although that was not true, especially as the Allies drove deeper gaps into the Gothic Line. It was a rumor, however, that the 10th Mountain Division didn't mind being spread.

Only after Hill 913 was securely occupied by Allied forces were the medical teams permitted to retrieve the killed and wounded. That's when the medics found me. One of the guys peeled my shirt back and pulled out my bloodstained dog tags, the metal name tags worn on a chain around a soldier's neck and imprinted with personal information, including name, rank, serial number, and blood type. (I still have my stained tags to this day.) Lifting my left arm slightly, he tied a six-inch label around my bloodied forearm, on which he wrote my name and serial number. On the tag were some preprinted boxes to describe my condition: the soldier checked WOUNDED; SEVERE; LYING (as opposed to a "walking case," or a "sitting case"). Someone pulled a litter—a stretcher with poles on each end that can be carried by two men or four—along the ground and slid it up next to me. Four guys, two on each side, placed their hands under me and lifted me onto the stretcher. They needn't have worried about jostling me; I couldn't feel anything below my chin anyhow.

The guys trudged down the steep slope of Hill 913, trying to carry me parallel to the ground, picking their way carefully across the pockmarked terrain, trying not to slip and fall on what was left of the slick, grassy field. At one point, the soldiers who were carrying me ducked below some trees, the stretcher scraped along the ground, and suddenly a sharp, searing pain shot up and down my spine. The pain was so intense I could feel it even though paralyzed. About a quarter of a mile away, at the bottom of the ravine, was an aid station. The stretcher bearers took me there for help, but the field medics took one look and quickly decided that my injuries were so severe that I needed to be taken to an evacuation hospital. They placed my litter on the ground along with a long line of other litters, all bearing wounded soldiers waiting to be transferred to the hospital. I don't know how long I lay there in my semiconscious state, but it was almost dark when a couple of soldiers picked up my litter and placed me on the back of an army jeep transport vehicle.

Sergeant Charles Dobbins of Yadkinville, North Carolina, was called upon to make the dangerous night run. Along with Herb Wolfe, a fellow medic, Dobbins loaded up the vehicle and took off for the forty-five-minute ride over the dark, narrow, treacherous mountain roads. By the time I finally arrived at the 15th Evacuation Hospital, it had been nine hours since the German shell shredded my shoulder and damaged my spinal cord. I had lost a lot of blood, so much so that the army medical staff decided to stabilize my condition before doing anything further.

I woke up in what looked to me like a large room with high ceilings and stark, bare walls. The light in my eyes made vision difficult, but it seemed as though there were at least a half dozen or more other beds and gurneys aligned throughout the room. I could barely move my head, but in my peripheral vision I thought I saw other soldiers stretched out on operating tables. I closed my eyes and drifted into semiconsciousness again.

Many of the soldiers from the 10th Mountain Division who were being brought in to the hospital were severely fatigued and dehydrated, and the doctors and staff recognized that it would be necessary to rehydrate them before attempting any serious operations. It was not until the following morning, at about ten o'clock, that Dr. Lyle French, a young neurosurgeon at the army hospital, operated on me.

Years later, Dr. French sent me his handwritten notes regarding my condition that day. He told me, "There is no question that you sustained a severe injury, not only of the bones but also of the nerves, in part from shell fragments and in part from the loss of blood because of the hematoma and the rupture of blood vessels."

Within twenty-four hours, Dr. French operated on ten men from the 10th Mountain Division's attack on Hill 913. In notes that he later recorded in his ledger of patients he treated that day, the doctor wrote about me:

Patient sustained penetrating high explosive shell fragment wound, entrance posterior aspect upper 1/3 right upper arm, traversed thru glenoid cavity up into right supraclavicular space. Occurred about 0900 hrs. 14 April '45. During the succeeding 24 hours, he developed a complete motor and sensory paralysis of his right hand and arm up to the axilla. His radial and ulnar pulsations were good. He developed a large hematoma throughout supraclavicular, submental and posterior scapula and axillary areas during this period. His general condition was good.

There was a compound comminuted fracture of the right humerus (upper 1/3), Right glenoid cavity, Right scapula and clavicle. 1st seen 1800 hours 14 April '45.

OPERATION: 1000 hours 15 April '45__General Oxygen Ether intratracheal anesthesia

The posterior wound was debrided up thru the glenoid cavity. The patient was then rotated and a long incision from base of neck, parallel to clavicle, over deltoid to about [the] intersection of deltoid was made. The fascias were incised, the clavicle elevated and the lateral cord and posterior cord of the brachial plexus visualized. There were numerous small bone fragments in and about both cords. These were removed with forceps and irrigation. The subclavian and axillary artery and vein were in continuity. There were several bleeding points, however, from which the hematoma developed. These were controlled with silk ligatures. The anterior wound was closed with silk. A thoracobrachial plaster cast was applied.

What that means in layman's terms is that Dr. French cleaned out the wound, then sliced me open from my neck out to the shoulder so he could examine the damage. Using forceps, he removed the numerous small bone fragments, and apparently used water to wash others away from the lateral and posterior cords. The doctor was not yet aware of my spinal injury. He tried to control my internal bleeding by sewing me up with silk stitches. When I was all stitched together, they applied a plaster cast, covering the upper half of my body, including my right arm, which was elevated at about a forty-five-degree angle.

It sounds painful, but, of course, I hardly felt a thing. Nor do I remember anything about that experience, except what I've been able to piece together from military and hospital records over the years. To this day, many of the details are a blur to me.*

In his report, Dr. French summed up his remarks regarding me: "Postoperative condition good. Evac to rear, 16 April '45."

No doubt, compared with many of the cases

---

* Many of my original hospital records were destroyed in a fire at the St. Louis Regional Records Center. As much as possible, they have been reconstituted and are now stored at the Robert J. Dole Institute for Political Studies, in Lawrence, Kansas.

that Dr. French encountered that day, I was in relatively good condition, but it sure didn't seem so. I was paralyzed from the neck down. Like so many other soldiers who had been wounded in combat, I had a few things going for me—I was young and strong, and had an incredible desire to live.

As Dr. French indicated, the following day I was transferred to Pistoia, a small town near Florence, where the army had taken over the local hospital. Built about the time that Christopher Columbus discovered America, the hospital had originally been a convent. Intricate mosaics depicting doctors and nuns assisting patients, as well as artistic scenes from the Bible, graced the eaves of the stone hospital's entrance. Inside, artistic renditions of Jesus healing the sick, crippled, and wounded covered the walls, some of the artworks stretching across an area of fifteen to twenty feet. The cupola windows, covered with dark paper to prevent notice by the bombers, gave way to high ceilings throughout the centuries-old building.*

* I visited the hospital as recently as November 2004, and the old section of it looks much the same today as it did in 1945.

Naturally, I knew none of this at the time. Heavily sedated, I was totally unaware of my surroundings or what was going on around me. Nor did I know that the consensus of the doctors who examined me was that my condition was much worse than originally thought, and I would probably die—soon. Whatever German explosive had hit me had damaged my spine as well as my shoulder and arm. With a prognosis of long-term disability, unable to move while lying on my back, I would be prone to pneumonia and all sorts of other problems.

The 15th Evacuation Hospital report said, "The patient had immediate paralysis, complete of all four extremities." A Captain Woolsey was a surgeon there, and he cut me open again to see if anything was pressing against the spinal cord, causing the problem. The doctors couldn't find enough to go on, so they simply sewed me back up and replaced the cast. There was nothing else they could do. I was done for, at least as far as ever having the normal use of my limbs. Whatever else was wrong with me, the docs were certain that I'd never walk again.

Of course, they didn't tell me that. Instead, Captain Woolsey and Colonel Prosser, the hospital chief, looked down at me and said, "You're going to be fine, son. You're just going to have to give it some time."

Time?

Give it time?

That's about all I could give it. I couldn't move anything other than my head and eyes. My upper body was in a new cast from my chin to my hips. My head was tipped backward, poking out of the body cast like a turtle's, with a sling under my chin attached to weights hanging over the back of the bed to keep my head from moving and causing further damage to my spinal column. My right arm was in traction. I had no control of my bodily functions.

Time? It lost all meaning as one day slipped into the next. The only good thing was that I was still alive.

Unaware that I had been wounded, yet trying to track events that had taken place near Castel d'Aiano, my dad wrote a brief letter to me that same week:

> **I just heard the [war] news, and it
> sounds really good today. We have
> been trying to keep in touch with your
> outfit, and know that you have been
> busy the past ten days. . . . Be sure**

and let us know what you need, and hope you can take another rest before long.

Business is good, and I think it will be [good] this spring as we have had a lot of rain.

So long for now. Write when you have time.

Dad

Several guys from I Company were in the hospital at Pistoia along with me. They were wounded but could still walk, so throughout the day, they'd stop by my bed to talk with me, adjust my pillows, and commiserate with me.

"You're gonna be okay, Lieutenant," they'd tell me over and over. "Just take it easy."

Under the circumstances, there was no other way I could take it.

Within a few days, it occurred to me that I needed to get a letter to Mom and Dad. No doubt, they'd be getting worried, not having heard from me since my letters from Rome when I'd been on leave. But how to write it, that was the problem. And what should I say? I didn't want to concern them unduly, yet I knew that eventually the War Department would be sending them a telegram informing them that I'd been wounded. I could just imagine James Wildman, the Western Union man who lived next door to us in Russell, coming to our house on Maple Street with the bad news. With a little luck, I could beat the bureaucrats to the punch, and save Mom and Dad some heartache.

One of the wounded guys from I Company offered to help, so I dictated a simple letter to him informing Mom and Dad that I'd been wounded in action, but assuring them that I was okay, that I was in a safe place and being well cared for. It would just take a little time, and I'd be fine. I have no idea what my buddy actually wrote to my parents. Apparently it was enough, yet not too much, because Mom responded promptly, bouncing back and forth between her motherly concern and her usual pluck and positive attitude, but not with any sense of panic or despair.

Dear Bob,
 **Daddy called to tell me that we had
a letter from you, and I knew
something was wrong—as I'd been**

following the 85th Infantry and 10th Mountain Division all over Italy. I was, needless to say, nervous and concerned about you, but I didn't think of what might have happened. But I know I was praying all the while that you were safe wherever you were. And I surely hope you are alright— but don't fail to be so very careful.

I'm sure you'll be alright, but I'm so sorry, but honestly, Bob, it's wonderful to know you're in a place where they are taking good care of you. So just rest and try to be as comfortable as you possibly can be. I'm anxious to know how it all happened and all, but of course I know you are handicapped, so I'm patiently waiting to hear. . . .

I only hope you get some more of the numerous boxes I'm sending. Only I hope you get the cookies and candy and not the chili I sent when it was so cold over there. I'm sure you've never received it yet, so goodness only knows what you might receive. I had sent fruit-cocktail three or four times. That might be better than chili now—also the cookies and letters and papers. . . .

Do you—I don't suppose you have any idea of how long you'll be there, do you? I only pray the thing will end before you get back, and don't let anything worry you. I imagine we'll be notified by the War department, but casualty messages are very short. [The] Banker [family] still hasn't had the official notice of Dean being a German prisoner and released, just the name in the paper as a Kansas boy. . . .

Tomorrow is the 25th of April, the Big Three Conference. I only wish Roosevelt could be there; I'd feel much safer over it all, because he's met with them before. But it may be alright.

Well, son, I'm so sorry, but I'm glad to know you're alright, that is, not any worse than you are. Daddy said he thought that you were very lucky, and so do I for a million reasons. But do tell us all about it. Hope you have

**something to read there, or cards to
pass away the time. But do take it
easy and let us know when you can.
Wish I could be there with you, but
maybe you'll do better this way. If you
want anything, please let me know.
Your letter came through in six days,
so it doesn't take long.**
  Good night, good luck,
    Love,
    Mom

Fruit cocktail, cookies, and all of Mom's
other goodies that she so faithfully sent to both
Kenny and me since we'd enlisted . . . sure
sounded good right now. How I'd eat them,
however, was another question. I still couldn't
sit up, my bowels and bladder were not working,
and I couldn't move. The nurses had IV tubes
hooked up to me to feed me, and a catheter to
relieve me. I passed the time staring at the dimly
lit beige ceiling of the old hospital walls, counting the squares in the ceiling pattern, thinking . . . thinking . . . reliving that day on Hill 913
as if in blurry slow motion, asking myself the
toughest question of all: **Why?** Why me? Why
did it happen? It was an unanswerable question,

so at the time I didn't bother trying. I'd have plenty of time to ponder the "why" questions later. Right now, I just knew that I was in bad shape. I was down, but I was not out. I had to get back up. After all, I wasn't the only guy in here who was hurting. Just look around: the room was filled with wounded men, many of whom were in far worse shape than I, struggling to survive. Many wouldn't.

As one hour slowly turned into the next, I concentrated my thoughts on trying to move my fingers, toes, arms, and legs. A strange paradox began taking place: As the sedatives wore off, the pain flooded in, throbbing intensely throughout my body. But as uncomfortable as it was, I almost welcomed it. The sheer fact that I could feel pain was an improvement; I hadn't been able to feel anything for days. Now the pain reminded me that I was alive, that my body was fighting with me to recover. I tried harder to move my fingers. I focused on my left arm, my left hand; I still wasn't feeling anything in my right arm, but I knew that portion of my body had been badly mangled, so I centered my efforts on my left side. Slowly, but surely, I could move my toes, and my legs just a little.

Move! I'd silently scream at my left hand and

arm, but it defied my orders. It lay motionless on my chest, atop the plaster cast.

Nevertheless, I remained upbeat. With the help of one of my buddies, I wrote home on April 25, 1945. Looking back now, I am amazed by my tendency toward understatement:

Dear Mom and Dad,

Just writing to let you know I'm feeling O.K. I can move my legs now, but I'm still having a little trouble with my left arm. I have a broken bone in my right arm and two in the shoulder. I guess some German thought I was a good target.

Write and let me know all the news. Tell me how Kenny is getting along in the Army. Tell everyone hello and to write. I'll let you hear from me as often as possible. Please don't worry about me. I may be home for my birthday [July 22nd].

Love,
Bob

CHAPTER 13

Inches and Miles

If you've ever had your arm or leg fall asleep because of poor blood circulation, you know that weird, funny feeling when that part of your body goes numb. Now, imagine that your entire body fell asleep for several weeks; then slowly you began to sense some feeling coming back. That's what I was experiencing during the last few days of my stay in Pistoia Hospital in Italy.

The tingle I felt in my legs and fingers, those barely perceptible sensations like tiny needles pricking my skin, motivated me to try even harder each day in my attempts to move. I focused as intently as I could for as long as I could on one particular part of my body, trying desperately to get a response. I'd grit my teeth until I thought I was going to crack them, as I attempted to **will** something to move, but nothing happened.

I couldn't raise my hands, I couldn't move my legs. Nothing. Then, ever so slowly, I noticed some feeling in my legs. I wasn't ready to go dancing, but I could wiggle my toes, and move my legs a bit.

Even more amazingly, on April 27, slightly less than two weeks since I'd been shot and instantly paralyzed, my right arm moved a few inches. The arm was in a cast, and I still couldn't feel anything in my right hand. Truth is, the motion may have been due to some reflexes contracting in my arm muscles, but it gave me hope, nonetheless. Overflowing with optimism, I sent word home that same day:

> **I'm feeling pretty good today. I'm just a little nervous and restless, but I'll be okay before long. I'm getting so I can move my right arm a little, and I can also move my legs. I seem to be improving every day and there isn't any reason why I shouldn't be as good as new before long.**
>
> **Send me something to read and something to eat.**
>
> **Love,**
> **Bob**

As I learned later, there were a number of reasons why I might never be as good as new . . . but I didn't know that then. And even if I had been aware of them, I would not have mentioned them to Mom and Dad. They had enough to worry about without being overly concerned about me.

Each morning I thought, **This might be the day good things start happening for me.** This could be the day: that I'll start getting better, that before long, the doctors will remove the plaster body cast, and I'll be on my way to recovery—to be able to breathe normally, to feed myself, to eat normally, to do the simple things in life we tend to take for granted, such as rolling over in bed or going to the bathroom; before long, I'll be able to walk, run, shoot a basketball, or catch a football just as well as I had done back in Kansas.

Just give me time.

As is often the case with any traumatic blow to a person's physical or emotional well-being, I didn't totally understand the seriousness of my injuries, and I was not ready to accept the fact that my life would be changed forever. On the morning of April 14, 1945, I could raise my right hand high in the air and motion the men in my platoon to follow me. It's been more than sixty years since that morning, and I've not raised my right hand over my head since.

To visit soldiers who have been injured, or anyone who is dealing with a disability that confines him or her to a hospital bed can be emotionally draining. But it's hard to overestimate how important and meaningful such visits can be. Some people avoid visiting someone who is incapacitated because they worry that they won't know what to say. Truth is, you probably don't need to say much of anything. You can be a tremendous encouragement to someone just by being there.

That's the kind of friend John Booth was to me. John was a young soldier from my platoon who came by to say hello almost every day during my hospitalization in Pistoia. Wounded in the foot, John was able to hobble along, so like many other guys, he went around the hospital helping where he could, encouraging those who would listen, and listening to those who just wanted to talk.

The best thing John did for me was to write to my parents. Along with my letter of April 27, John decided to add a note of his own. He wrote:

Hello, Mr. and Mrs. Dole:

I'm sure you know that Robert is unable to write so I tried to write him a note. He told me what to write. I know you are worrying about Robert but I wouldn't worry too much because there isn't any doubt in my mind at all but [that] he will be just as good a man when he gets well as he was before he was hurt.

Just thank God it wasn't any worse than it was. That's the way I feel about it. In case you want to know who I am, my name is John Booth of Bethany, Mo. Robert was my Platoon Leader. He is a fine fellow. I'll write again for him. (A sniper shot me in the foot. I can't walk very well but it won't be long until I can.)

As always,
John

That same day we learned that the Russian Army had broken through the outer defenses of Berlin the day before, on April 26, 1945. Simultaneously, the Americans and Allied forces were knocking holes in the remnants of German resistance to the west of the city. It was finally happening. The Third Reich was being ripped apart, block by block, building by building, brick by brick. A few days later, in the early-morning hours of April 30, Adolf Hitler committed suicide.

I was still in the evacuation hospital in Pistoia on May 2, 1945, when what was left of the German Army surrendered in Italy. The war was over—at least our part of it—although the official surrender in Europe didn't take place until six days later, on May 8, 1945. A shout went up, resonating through the wards of the quiet Pistoia hospital, when the news came through that victory was ours.

That same week, Jimmy Wildman, the Western Union operator in Russell, received a telegram to deliver to my mom and dad. Ironically, Jimmy Wildman had often solicited the help of my dad to deliver the War Department's sad messages to other families whose sons were casualties in the war. Dad had a comforting way about him. Even Reverend Jenkins, the pastor of our church, often asked Dad to sit with family members who were sick or who had lost a loved one.

This was one telegram, however, that Jimmy

Wildman would have to deliver personally. The telegram, sent on May 3, 1945, at 8:19 A.M., was, as Mom had predicted in her letter to me, short and to the point.

> The Secretary of War desires me to express his deep regret that your son, Second Lieutenant Dole, Robert J., was seriously wounded in Italy, 14 April, 1945. Hospital is sending you new address and further information. Unless such new address is received, address mail to him: Write name, serial number, Hospitalized 2628 Hospital Section APO 698 c/o Postmaster, New York, New York.
>   J.A. Ulio, Adjutant General

That was it. For all its good points, the army could be extremely cold and slow at times. No words of consolation were included in the telegram. No wishes for a speedy recovery. No details regarding how I was wounded, or how badly, or where I was being treated. Just a blunt message that I had been seriously wounded.

Fortunately, Mom and Dad had been emotionally prepared for the telegram a week or so earlier, thanks to John Booth's writing my dictated letters. Pity the many parents, then and now, who are recipients of such a jolting message from out of the blue.

The battle in the Pacific continued to rage, although everyone knew that once the Allies were able to turn their full attention to the east, the Japanese—already weakened by General Douglas MacArthur's forces—wouldn't stand a chance. Still, I was concerned about Kenny. I knew he was over there somewhere. Just as I had gone unscathed throughout the entire war, only to be wounded three weeks before the Germans surrendered, I recognized that Kenny would be in danger until the day the Japanese capitulated as well.

The doctors at Pistoia had done their best, and once my condition had stabilized, they sent me to a bigger hospital in the Mediterranean Theater, the 70th General Hospital, in Casablanca. Movie stars Humphrey Bogart and Ingrid Bergman had made the city famous for millions of Americans when they starred in a story of reignited love and unlikely heroism on the big screen in 1942. But for me Casablanca was just a stopover.

A few days before they shipped me out to Casablanca, Captain Woolsey and Colonel Prosser were making their rounds in the ward. Pausing beside my bed, Captain Woolsey said, "Go ahead, Lieutenant, show the colonel what you can do."

I forced a partial smile and went to work, attempting to move my left arm. I struggled and strained for nearly a minute while the doctors waited and watched patiently.

Nothing.

Then, as though I'd received a sudden influx of strength from on high, I lifted my left arm a few inches off my chest. You'd have thought I'd set a new world's record in weight lifting. The doctors and the other patients were all smiles. It was only a few inches, but it was miles up the road from where I'd been three weeks earlier. Maybe there was hope for me after all.

Despite my much-heralded achievement, on May 16, 1945, I was prepared for the flight to Casablanca. It was unnerving to be so helpless. Picture yourself being unable to move, your upper body encased in plaster, your head strapped back to prevent it from rolling or from snapping forward, your one free but useless arm strapped down to your body to keep it from falling. But they were taking me out—taking me closer to home.

At the airstrip, another group of men lifted me from the ambulance. The springtime breezes ruffled the sheet covering my cast, as the men hoisted me aboard the hospital plane and strapped me in using a series of belts hanging from the plane's walls and ceiling. I was given another sedative prior to my departure, and the next thing I knew, I was waking up in a hospital ward at the 70th General Hospital, in Casablanca.

Rumors circulated from the start that the doctors at Casablanca didn't hold out much hope for my recovery. Almost as soon as I arrived there, they began talking about expediting my flight home to the States. The army had an unwritten policy: if a wounded soldier looked as though he might not survive, he should be sent home as quickly as possible. It would be much easier on everyone involved. I didn't care. The mere thought of being back on American soil sent electricity through me—at least, I think it did. With my lack of feeling, it was hard to tell. One thing I knew for sure, however: my heart leapt at the possibility of going home. As soon as

I found someone to help me write a letter, I sent a note to my parents.

> **Dear Mom and Dad,**
>   **I haven't written in a couple of weeks, primarily because I thought I'd be home by this time. I'm in a different hospital now and I should be going home soon. Am feeling much better than I was when my last letter was written. My legs are better and my left arm seems to be improving steadily.**
>   **The cast I'm in is none too comfortable but as soon as I reach home, it will be taken off. There is a possibility that I will be sent to Winter General in Topeka.**
>   **Love,**
>   **Bob**

Conspicuously absent from my letter, of course, was any mention of my shattered right arm and spinal injury. Part of that was due to the fact that my broken bones were still healing, and part of it was due to the fact that I could

barely move my fingers, and had almost no feeling in my right hand. I didn't want to worry my
parents any more than necessary.

Meanwhile, life in the Casablanca hospital
was pretty much the same as it had been in Pistoia. I spent most of my days and nights lying on
my back, waiting for someone to come by to
talk, to feed me, to help me write a letter, or to
help me smoke a cigarette.

Smoking was a newly acquired habit. I had
never really smoked prior to entering the army.
Oh, sure, I'd tried a few puffs at the fraternity
house while in college, but at KU, I was an athlete. I'd always been conscientious about taking
care of my body, while others may have lit up in
naïve ignorance of the dangers of nicotine.

Life in the army made many of us into smok-
ers. In every package of "C" rations, I found
some stew that had to be heated over a fire,
some crackers and cookies, and at the bottom a
package containing four cigarettes. At first, I
wasn't too interested in smoking the cigarettes,
so I'd give them away to my friends. But after
getting the cigarettes day after day, week after
week, I finally thought, **I think I'll try these
things. They must be okay. After all, the
army is giving them to me every day. If**

the cigarettes weren't good for us, the army wouldn't put them in our food containers. . . .

I lit up a cigarette, sat back and thought, **Boy, what a life!**

America's tobacco companies had discovered a built-in market: the U.S. military. Cigarettes soon became a regular part of the military culture. A cigarette after a meal was practically like dessert. And I'd always loved dessert. Before long, I was smoking like a chimney.

Interestingly, my addiction didn't disappear just because I'd suffered a devastating blow to my body. Quite the contrary. Lying in a hospital bed in Casablanca, unable to move, I craved a cigarette all the more. Because smoking was an opportunity to talk with one of my buddies for a while, I'd often ask someone to hold the cigarette for me.

The most immediate problem with smoking, though, was flicking the ashes. Usually the person who lit the cigarette for me would stick around while I smoked it. But sometimes a nurse or soldier would light the cigarette and then walk away. Ordinarily, most people who smoke don't allow their cigarettes to burn all the way down; they tap the ashes into an ashtray.

But with my arms and hands not working, I had to flick the ashes using my lips. At least once, I flicked the hot ashes right down my neck and inside my cast. It was probably good that I wasn't feeling much during those days. Beyond that, it's a wonder I didn't set my bed on fire.

Cigarette ashes weren't the only things that found their way down inside my cast. I couldn't sit up completely to eat, so even though someone had to feed me, I'd sometimes lose part of the food down there as well. Everything from vegetable soup to bread crumbs seemed to find its way between the plaster and my skin. There may even have been some bugs in there having lunch, for all I knew. Besides causing my flesh to be itchy, with no way to be scratched, the cast began to smell awful. Each day, it got worse. Just thinking about the fact that parts of my body had not been washed in more than a month made my skin crawl. My fastidious mother would have had a fit.

I remained at the 70th General Hospital throughout the month of May and into the early part of June. The doctors and nurses tried to make me as comfortable as possible. There wasn't much more they could do. One day dragged into the next.

During the first week of June, nearly two full years to the day since I'd been called up to active duty, I received the best news I'd heard in a long time—the army was shipping me back to Kansas. I was going home.

CHAPTER 14

Homeward Bound

The poet Robert Frost once defined home as "the place where, when you have to go there, they have to take you in." Maybe so, but there's no better feeling than knowing you are going home, especially when you've been away for quite a while.

On June 3, 1945, the doctors in Casablanca examined me one more time, and then began preparing me to head back to the United States. Once again, I was crated up like a piece of furniture to make the long trip from Casablanca to Miami. Nevertheless, my spirits were already soaring on June 6, 1945, when I was transferred by litter to an ambulance that took me to the airfield, where I was lifted aboard a giant aircraft.

Inside, the airplane bore little resemblance to a passenger jet; there were no seats to speak of;

instead it was a cargo plane set up like a field hospital, complete with doctors, nurses, and medical equipment. Still lying on my litter, I was hoisted onto the upper deck of two tiers of hooks; I couldn't tell for sure, but it looked as though other soldiers in their litters were being "stacked" on hooks below me, hammock style, as if we were consigned to double-decker bunk beds. Dozens of men, and possibly a few women—I couldn't really see—lay in similar states of limbo. We remained that way throughout the flight, with the nurses coming by occasionally to check on us. It was the strangest flight I would ever take in my life, yet one of the best. A nurse came by and gave me a shot. "Don't worry, Lieutenant. You'll be okay." The shot must have been powerful, since it soon knocked me out. I was barely conscious, but I was on my way home. Any discomfort the other soldiers or I may have felt was overcome by the anticipation of seeing our friends and families, and resuming the lives we had left behind. We arrived in Miami the following day, and I was transferred to the Army Air Force Regional and Convalescent Hospital in Miami Beach.

I had been out of the country only a little more than six months, yet it seemed a lifetime since I had boarded that troopship and set out from Fort Meade, Maryland, for Naples, Italy. So much had happened, and I was thrilled to be back. Had I been able to, I'd have gotten down on my knees and kissed the ground. Instead, as soon as possible, I asked permission to telephone my mom and dad. A fellow wounded soldier held the telephone receiver for me as I assured my parents that I'd be home soon.

"I think they are transferring me to Winter General Hospital in Topeka," I told my mom. I had barely gotten the words out of my mouth before Mom was making travel plans to meet me in Topeka. My dear mother had no idea what she would see when she arrived.

I had barely been admitted to the hospital in Miami Beach when, on June 11, I was packed up once more for shipment to Winter General Hospital, in Topeka, Kansas. Winter General was an old Quonset-hut type of facility that had been activated as an army hospital in December 1942. At its peak, it once housed 2,200 beds. With the help of the renowned psychiatrist Dr. Karl A. Menninger, Winter General became known as a psychiatric teaching center and was transferred to the Veterans Administration in December 1945. It became the first army hospital in America to be taken over and run by the VA following the war. Thanks to Dr. Menninger's leadership, the facility became a model for other VA hospitals dealing with emotional and mental traumas, as well as the physical problems that GIs brought home from the war. Totally renovated, today the facility encompasses more than twenty buildings, and is known as Colmery-O'Neil VA Medical Center, named for two outstanding Kansas veteran leaders.

When I was admitted to Winter General, on June 12, 1945, I couldn't have cared less about the hospital's storied past or its bright future. I was just glad to be close to home, and a mere twenty miles or so from the University of Kansas campus. Being that close, I couldn't help dreaming of the day when I'd be back playing basketball for Coach Phog Allen, playing football, and running track at KU once again.

Mom traveled by herself from our home in Russell—178 miles away—to be at Winter General when I arrived still encased in the plaster cast from my ears down to my hips, the same cast that the doctors had put on me in Italy. I must have been a frightful sight for my mom. Appearances aside, I was running a nagging fever that seemed to be edging higher and higher. My weight had been steadily dropping from the time I had been wounded. No doubt I looked frail and fragile, a far cry from the strong, strapping athlete who had gone off to war two years earlier.

Looking back, I'm sure the nurses at Winter General tried to prepare my mom for our first encounter, warning her that besides being wounded, I'd been sick, that I couldn't move, and that neither arm was functioning. But it was useless. No matter how well they attempted to steel my mom, it wouldn't have been enough. The moment she stepped in my room and caught sight of me, she gasped. Stopping in the doorway, she stared in silence a moment and then burst into tears.

I'd only seen my mother weep a handful of times in my entire life. I couldn't recall ever having caused her to cry by something that I'd said or done. Now, as I tried my best to roll my head slightly in her direction so I could better see her face, I desperately wanted to console her, but I could not speak.

Then, an unusual transformation took place. It was almost as if Mom said to herself, **Okay,**

Bina. That's enough. You must be strong. She took a clean white handkerchief from her purse, dabbed her eyes, and walked over to my bedside. She reached out to me, touching my cheek softly and stroking my face. "Oh, Bob," she said, the tears welling again. "Oh, Bob . . . oh, son, I'm so glad you're home." She may have wept bucketfuls outside my hospital room, but Mom never again cried in front of me.

She sat down next to my bed and stayed there all day long, and all night, too. I'd nod off from time to time, but every time I opened my eyes, she'd be sitting there, watching. That is, when she wasn't cleaning me up. One of the first things Mom noticed after our initial reunion was the stench emanating from inside my cast. With all the food droppings, cigarette ashes, and who knew what else, it smelled as though something had died down inside there. Maybe something had.

Mom scrubbed every inch of me that was exposed, including the exterior of the cast and as far as she could reach her fingers inside it without hurting me. Whether she'd previously made up her mind to do so or not, I don't know, but Mom decided to rent an apartment across the street, within walking distance of the hospital. She refused to leave my side any more than absolutely necessary.

When I think about the sacrifices my parents made for me during those days, it deeply moves me and is extremely humbling. My parents were blue-collar-type folks. In 1945, Dad was still working at the grain elevator in Russell; Mom was selling sewing machines and vacuum cleaners. Even before the war, they had rented out the upstairs of our house to help make ends meet. Now, after four years of hardship, when it looked as though things might finally start turning around for them financially, they selflessly drained their meager savings so Mom could stay with me in Topeka.

At night, when she finally tore herself away from my room, she'd go to her apartment and call my dad, telling him about our day, how she had fed me, bathed me, or had done anything else I wanted. To Mom, I was almost like her baby again, needing constant care.

Horrifying my mother nearly as much as my appearance, was my recently acquired habit of smoking cigarettes. Mom detested smoking, even though my dad had smoked like a human smokestack for years. She was always going

around the house cleaning up his ashtrays, or reminding him to throw out his cigarette butts.
She actually preferred that Dad smoke on the
porch, although she never said that. She wasn't
bashful about letting him know what she
thought about his filthy habit, though.

You can imagine, then, what an expression of
unconditional love it was for my mother to hold
the cigarettes up to my lips while I smoked. She
never flinched as clouds of blue smoke swirled
around her head. Instead, she acted as though
the smoke didn't bother her a bit, though I knew
it did.

But she loved me that much. She didn't want
to deprive me of one of the few physical plea-
sures I could enjoy. It would be more than
twenty years later before I'd break the smoking
habit. My dad and brother never did.

One day, while Mom was holding a lit cigarette for me, it slipped out of her hand and
dropped down inside my cast—still burning.
With my impaired sense of feeling, I hardly noticed, but my mother nearly had a heart attack.
She did the first thing that came to mind. She
grabbed a glass of water sitting nearby and
dumped it down the front of my cast. You can
imagine how that smelled after a few days.

The doctors at Winter General put me through an extensive neurological examination. They didn't seem too worried about the fact that much of my shoulder had been blown to bits. They were far more concerned about the damage to my nerves and spinal cord areas. According to my army medical records, they eventually diagnosed the injuries to my "cervical cord near [my] C-4 vertebrae, as well as injuries to the backside roots of the C-3 to 8 vertebrae, causing the complete paralysis of my upper right side, with partial paralysis and weakness of my left upper body, resulting in only minimal sensory impairment." In other words, I could feel a little on my left side with hopes of gaining further movement, but the upper right side was a mess.

Still in the cast, I spent most of each day straining to move my fingers, arms, and legs. More feeling was coming back to my legs now, a little more each day, and my left arm, too, although the thumb, index finger, and middle finger remained numb. To this day, one of the best-kept secrets regarding my war wounds has been the damage to my left arm and hand. Over the years, I've shaken hands with so many people using my left hand that most people have assumed that the hand is fine. Actually, I have no more feeling in those fingers today than I did in June 1945, and the left hand is extremely sensitive. After shaking hands with a few too many folks, my left hand starts turning black and blue, much as it did sixty years ago.

In addition to the doctors and nurses who worked with me, Winter General was blessed with a bevy of wonderful Red Cross workers, "the Gray Ladies," as they were known. The American Red Cross was founded in 1893, and the Topeka chapter was one of the first local groups, dating to 1910. By the time Winter General became an army hospital, much of the nonmedical help was done by the Gray Ladies, volunteers who dressed in gray uniforms and served faithfully without reward or recognition. Their work ranged from visiting the sick and wounded, to serving at information desks, writing letters, reading, tutoring, and shopping for patients who needed items unavailable at the hospital. The Gray Ladies came by almost every day to cheer me up, to massage my hands, or help exercise my legs and fingers, to help the nursing staff in every way possible.

Each day became a marathon of endless hours trying to exercise my legs, my left arm, and the fingers on my left hand, with Mom and the nurses cheering every small triumph. On good days, I could move a finger or arm a little; on bad days, I struggled to move at all. I felt imprisoned in my frozen body. I still could not control my bladder or bowels; nor could I sit up in bed. It was a monotonous existence. Nobody said much at all about my right arm, which had no discernible feeling.

Mom spent every day at the hospital with me. Dad and Norma Jean drove the nearly three-hundred-sixty-mile round trip every weekend. Having the family close to me was a mixed blessing. On the one hand, I was so glad to see them; on the other, I hated having them see me so helpless.

Often after I'd tried for hours to move my arms, Mom or Norma Jean would suddenly hurry out of the room. A few minutes later, they'd return, their eyes red and puffy. Sometimes after Dad and Norma Jean said good-bye, and Mom returned to her apartment for the night, I'd lie in bed staring into the darkness, asking myself again and again, **Why? Why me? Why was I on that Hill 913? What did I ever do to deserve this? Why wasn't Somebody up there looking out for me?**

In my better moments I realized, **Some-body was.**

The discomfort of my circumstances was enough, but what really shattered me was the frustration of not being able to move, sit, stand, or eat normally. I was totally dependent, a basket case, barely able to function. Dark waves of self-pity enveloped me, threatening emotional as well as physical collapse.

I usually got over these bouts with despair before my mom returned to the hospital the next morning, but not always. Once, for instance, my aunt Mildred drove all the way to Topeka to see me, bringing a banana cream pie, my favorite. But I was discouraged and didn't feel like talking.

"I brought you a banana cream pie, Bob," Aunt Mildred said as she leaned over my cast and kissed my cheek. "Just the way you like it."

"Good," I said. That was all. My response was uncharacteristic, and I was ashamed of it, but I just didn't feel like talking.

Sometimes Mom helped pass the time by reading books or the newspapers to me, or relating the latest rumors around Russell that she'd learned from Dad by phone, but now she was much more careful to "edit" the news, shielding me from information she thought would worry or discourage me. Kenny was still in New Guinea, in the Pacific, as was my good friend Bud Smith. We'd heard that Bud had been listed as missing in action, and that my college fraternity buddy Herb Finney had been shot down somewhere in the Pacific. That didn't sound good, but we were hoping for the best. Herb's mom had been such a good friend to my mom, often staying with her at the hospital. Then, one day in late June, Mr. Finney called. He'd received a telegram from the War Department.

I didn't see Mrs. Finney for a long time after that.

Fever had become a perpetual part of my existence since returning to the States, and it flared up during the latter part of June and the first few days in July. The higher my temperature rose, the less coherent I became. At times, I felt myself slipping away, and I'd try desperately to hang on to consciousness, hearing my mom's voice calling me . . . "Bob. **Bob.** Son, I'm right here with you." Then the darkness took over.

On July 10, the fever intensified, my temperature creeping up to a dangerous level and beyond. Unable to control it, the doctors told my mom to prepare for the worst.

Mom called my dad at work. "Doran, Bob is in bad shape. Get Norma Jean and come as soon as you can. The doctors say he might not make it through the night."

Somehow the Russell police got word that I was in trouble. They called ahead to the Kansas State Police, who provided Dad and Norma Jean with a siren-screaming police escort the length of Highway 40, all the way from Russell to Topeka.

By the time Dad and Norma Jean arrived, my temperature had soared to a scary 108.7 degrees (axillary). My loved ones watched in silence as doctors and nurses wrapped my body in a rubber sheet and then frantically started packing ice around me, like a fish. It was their last hope . . . and my last chance.

They had figured out by now that my right kidney was terribly infected and contained a large number of stones. The doctors wanted to remove the kidney, but they didn't dare as long as I ran the sky-high fever. In desperation, they decided to inject me with the new miracle drug of the day—penicillin—in hopes that I could hang on long enough to let the combination of the drug and the ice bring my temperature down to a point where they could operate.

Mom prayed. Dad paced. Norma Jean fought back her tears and tried to hope. All they could do now was wait.

CHAPTER 15

Who Is That Man?

Throughout the night of July 10, 1945, I faded in and out of consciousness. The fever racked my body, intermittently causing me to perspire profusely one minute, and the next chilling me to the bone. Even being sedated and with my diminished sense of feeling, I was aware of the cold ice around my entire body. Fans were blowing on me from everywhere, but I felt like I was still burning up.

Nowadays, any time I forget something, or lose my train of thought in a speech, I might wisecrack, "Sorry, I can't remember. My brain was cooked by high body temperatures in July 1945." The remark elicits smiles from those who know me well.

But it was no laughing matter that July night in Topeka. Doctors at Winter General monitored my condition all night, along with Mom, Dad, and Norma Jean. Somebody later told me that a chaplain stayed nearby.

Finally, the fever broke. Ever so slowly my temperature dropped. Further. A little more. Close enough. The next day, July 11, the doctors acted quickly, rushing me into surgery. They removed my right kidney, and it was over before I knew what had happened.

Once the infected kidney was removed, my body began to respond to treatment. Since April 14, when I was wounded, to July 11, when my kidney was taken out, my weight had dropped from 194 pounds to 130 pounds, and was still falling. I felt weak and scrawny, and sore from lying in bed for so long.

Nevertheless, by late August 1945, I had recovered the use of my bladder and bowels. I could also move my legs, and I'd gained some feeling in my left arm and hand. I still couldn't sit up, get out of bed, or go to the bathroom alone, but at least I was free from the catheter.

Motivating me even more, one day Mom told me that I had some visitors. I didn't want visitors; I didn't want to see anyone and didn't want anyone to see me in my condition, and besides, it was downright difficult to carry on a conversation lying on my back, barely able to move my head to see somebody standing above me. But when Mom told me that two special friends had come to see me, I said okay.

Into my hospital room walked none other than my KU basketball coach, Phog Allen, and the KU trainer, Dean Nesmith. I could hardly believe my eyes.

Phog Allen, the venerable coach with the foghorn voice, didn't know how to talk quietly, and probably wouldn't have if he could have. Even patients and hospital staff who weren't big KU basketball fans couldn't help noticing him. He seemed to have such charisma, a real presence. He stepped up to my bedside and laid a hand on my left shoulder. He was amazingly upbeat and positive in his conversation. It was as though he didn't even notice the fact that my right arm and half my body were encased in plaster, and my left arm and hand barely functioned. No, Coach Allen immediately started talking about how good it was going to be when I got well, and back to KU.

Dean Nesmith, on the other hand, looked troubled. Not that he was disagreeing at all with Coach Allen, but more as though he might be recalling our last meeting—back in Rome in

early April, when we were talking about the possibility of my attending the Sparta School, the
army's sports program, for two weeks. Had that
happened, I may have been in Rome on April
14, 1945, rather than on Hill 913.

Letters of encouragement poured in from all
sorts of people. Everyone from Reverend Jenkins to friends I'd known back in Russell High
School, but hadn't seen in years, had written,
and Mom read every letter to me. Sometimes, if
I was feeling okay, I'd dictate responses to the
folks who wrote me, through Mom or one of the
Gray Ladies. While some of the letters may have
gone unanswered, they were all appreciated.

Even my dear old grandmother got into the
letter-writing routine. Grandma Dole wrote:

> **Our dear grand son,**
>   **Now Bob, I must tell you how happy
> we were to know you are feeling so
> much better. Why it won't be long
> until you will be home. Won't that be
> grand . . . wish you could have been
> with us Sunday. . . . Yes, we had ice
> cream. No potato salad or baked
> beans. We will have a victory dinner
> one of these days with you and**

Kenneth both with us again. . . . If
your mother stays away much longer,
she won't want to come back to little
old Russell. Anyway, it's nice she can
be with you. Norma is sure a lovely
girl to work and keep house, and
always the same. Really hope you are
enjoying this nice cool day. Saturday
was awful hot on anyone, say nothing
about you boys there in the hospital.
Didn't you nearly roast? Wonder if
Kenny won't get to come home soon. I
had better close, as I must churn, or
no butter for supper.

Now Bob, be careful. Don't try to
get up and walk home. It's just too
far. You might play out before you got
here. Ha. Lots of love, and hope every
day you will be stronger.

Your Grand Ma

And people wonder where I got my sense of
humor? I hadn't set foot on a floor in six
months, and Grandma was warning me not to
try to walk the 178 miles home to Russell.

But Grandma's letter served to remind me of
some things—mostly that she believed in me.

The army doctors in Pistoia, Italy, thought that I'd never walk again. The physicians at Winter General thought that I'd never walk again. But if Grandma thought I could walk again, I was certainly going to try.

It was far from easy, though. The nurses began by helping me to sit up. Some days, that's all I could do before I'd be exhausted. Then, slowly, they slid my legs over to the side of the bed and let them dangle off the side. Just having my feet pointed in the right direction was a whole new experience. But fatigue set in quickly. Moreover, almost any movement caused me to shake with tremors and spasms. The doctors at Winter General noted in my army medical records that many of my nerves and muscles were operating at only about fifty percent. Others were weaker yet.

It took days for Mom and the nurses to get my legs all the way off the side of the mattress to where they touched the floor—mostly because I was afraid of falling. "Take it slow, now," they'd say, as they allowed a little more of my weight to be supported by my legs and feet each day. Then one day, with a nurse on each side and one ready to jump in behind me, they slid me all the way off the bed, until my feet were firmly on the floor. "Easy now. Let's go a little more." I could feel my backside being lifted up off the bed until I was standing on my own legs.

For a long moment it was incredible. **I'm standing,** I thought. The last time I had stood on my own two feet was the morning of April 14, 1945. Slowly, ever so slowly, and only for a brief moment, the nurses took their hands off me. I was standing up all by myself.

But then came the tremors. I started to shake violently. The spasms attacked my body. I tried as hard as I could to make the involuntary movements stop, but my legs kept on shaking. It wasn't just my legs, though; my entire body seemed on the verge of convulsions. I teetered, tipped forward, and for a moment I felt certain that I was going to fall flat on my face, with my arms unable to reach out to break my fall. Fortunately, the nurses were quick to pull me back onto the bed, easing me down into a prone position. I'd lie there for hours, exulting in the fact that I had stood for at least a second or two, discouraged that I hadn't been able to stand for longer, and totally exhausted from the physical and emotional effort.

It was much the same routine every day. They'd slide me out of bed, stand me up like a scarecrow dressed in a plaster cast, and I'd stand there, trying to will my feet to move; I'd stand for as long as I could until the tremors forced me down again.

Sometimes when I was shaking uncontrollably, Mom would have to leave the room. She didn't stay gone long, though. A few minutes later, she was back, her eyes slightly red, but with a smile on her face and an encouraging word on her tongue. "You're doing fine, Bob. Keep going. You can do it. You're going to be walking all over this hospital soon."

I wasn't sure I believed her, but I wanted to.

One day, the attendants got me up, and after standing for a few minutes, working up every ounce of willpower within me, I felt it. My left foot moved. Not far. Not even an entire step; it was more like inches, I don't know, but I knew that I had moved under my own power, without the assistance of a nurse or orderly.

The doctors decided to replace the plaster cast with a less cumbersome one. I was ecstatic. I felt like a new man with that heavy, dirty hunk of plaster off my chest and back. The cast had served an important purpose, I guess, keeping my head and spine straight and allowing some of the injury to my back and neck to heal. But I was thrilled to be released from that plaster prison. All the more, I wanted to get up and walk, now that I was free of that albatross.

Then one day, as Mom and the nurses were helping me out of bed, I looked up and noticed that somebody had left the bathroom door open. I could see a mirror in front of me, but I didn't recognize the image in the glass. The last time I'd seen my reflection in a mirror, I weighed 194 pounds. But by now, my weight had dropped to about 122 pounds; my arms and legs were so thin I looked as though I'd been in a prisoner-of-war camp; my formerly neatly combed, thick black hair was mussed. My strong upper body, once finely chiseled and toned by my lifting the concrete-block weights, now looked puny and concaved. My legs had atrophied to the point that they looked like a crane's legs sticking out from under my pajama top. My eyes were sunk so far back in my head I looked like a ghost. My right arm stuck out in a new triangular-shaped brace, and my left arm dangled awkwardly. My hands were shriveled and gnarled. When I saw myself in the mirror, I was horrified. It was the first good look I'd had of myself since before that morning outside Castel d'Aiano, back on April 14. **Who is that?** I thought. **That can't be me.**

It's been more than sixty years since I first saw that image in the bathroom mirror. In the past sixty years, I've glanced at my full body in a mirror less than half a dozen times. Except to shave and comb my hair, I still avoid looking in mirrors. After showering in the morning, the first thing I do is put on a T-shirt. I don't need any more reminders.

Every day, the nurses or attendants would get me out of bed, and I'd struggle to take a step. First a small shuffle, then an actual step. Sometimes I could take several steps before getting tired. It was painstakingly slow, but I was determined, and I was making incremental progress. After a while, I attempted to shuffle across the room without assistance, moving from one stable piece of furniture to another, one wall to another, just in case I needed something to help me keep my balance when the tremors struck.

One day, the doctor was standing at the bottom of my metal hospital bed scribbling some notes on my charts. He looked up at me and said some of the best-sounding words I'd ever heard. "Lieutenant Dole," he said, peering over my chart. "How'd you like to go on leave?"

"Do you mean . . ."

"Yes, that's what I mean. How would you like to go home for a while?"

"Are you serious?"

"Yes, of course. There's not much more that we can do for you here that you couldn't do for yourself at home. Might be good for you to get some home cooking for a while." He looked at my mom and smiled, then turned back to me. "You'll have to take it easy, though. Keep up your exercises, but don't push anything."

Although I was a patient in Winter General Hospital, I was still a soldier in the United States Army. The doctor authorized a thirty-day leave, beginning September 12, with a possible seven-day extension, which meant I could stay home until October 19.

I could barely believe what I'd heard. My recovery was far from complete. I couldn't feed myself or bathe myself. In truth, I could hardly move. I couldn't move my right arm at all, and I could manipulate my left hand only a little, but I wasn't going to pass up an opportunity to go home.

Mom called Dad to come for us, and early on September 12, 1945, the hospital attendants helped them load me into the backseat of the car. Dad fired up the motor and eased the car onto Highway 40, westbound, in the direction of Russell, Kansas. I was going home—home to Russell, home, where I belonged.

CHAPTER 16

Leaving on My Mind

I had never appreciated my hometown as much as I did that day in September 1945 when Dad drove down Main Street so I could see it again, past Dawson's Drugstore, past the grain elevator where he still worked; we crossed the railroad tracks and turned left up to our house on the corner, at 1045 Maple Street. The house was just as I remembered it, although it looked as though Dad had been working hard in the front yard. As always, every tree was trimmed perfectly, every blade of grass cut as though Dad had done it with fingernail scissors. The cooler fall temperatures were already causing some of the leaves on the trees in our yard to turn colors. The flowers on the rose trellises on both sides of our front porch had lost their bloom, but the greenery was still visible.

I was concerned about getting into the house. Although our front yard was flat, two vertically laid brick steps led up to the front porch. But I could not raise my legs that high. My brother, Kenny, was now stationed in Cebu, in the Philippines, and my sister Gloria—who had come to visit me in Topeka a few weeks earlier—was back in Lincoln, Nebraska, waiting for her husband, Larry, to get out of the air force. It was up to Norma Jean, Mom, and Dad to get me inside the house. I'm afraid I wasn't much help. I was like a sack of potatoes, dead weight in the backseat. It took forever to slide me out of the vehicle and onto the stretcher. Then Dad, Mom, and Norma Jean carried me up the steps and inside the house.

Of course, in a small town like Russell, with as many friends as Dad and Mom had, they could have had a dozen strong men on hand to help carry me into the house. But they knew that I wouldn't have wanted anyone but family to see me right now. It wasn't false pride; it wasn't that I was so ashamed of my appearance, although I certainly was. It was more than that. It was my inability to do anything, my sense of helplessness that I wanted to hide from public view. I had been a big, strong kid in Russell when I'd gone away, the guy everyone in town remembered running through the streets early each morning. I had been a leader on the football, basketball, and track teams. Now I was a 122-pound weakling who had to be carried into the house, helped to the bathroom, and fed like a baby.

By the time the family got me inside the house we were all exhausted. The last twenty feet of the trip from Castel d'Aiano to our living room in Russell had been some of the toughest. But I was happy to be home.

Mom and Dad had rented a hospital bed and put it in their bedroom, just off the living room, separated by a set of French doors. They moved their belongings to the small back bedroom, and gave me their front bedroom, so I could be close to everyone. With the French doors open, I would be able to see and hear all the activity out in the center of the house.

A day or so after I got home, I received a letter from Kenny. The letter had been sent to Winter General and forwarded on to me at home. Kenny's note made me smile.

Bob,
I do hope that you will receive this short letter at home, and not in Ward

18 [at Winter General], because I know it is just what you need to put you back on the ball—and if this does find you at home, please eat about two pounds of fresh fried liver for me and everything else that Mom fixes to go with it, about a gallon of home made ice cream, and two butterscotch pies, and anything else you can get your hands on, and when I get home I will help you out. . . . Take it easy and get well soon.

> Love,
> Kenny

Kenny's menu was just what the doctor ordered. From the moment we got in the door, I don't think Mom ever stopped cooking. She was always making something for me to eat—especially liver and onions, one of my favorite dishes—trying to fatten me up, to get some meat on my bones. She worked constantly to take care of me, bathing me, dressing me, combing my hair, helping me to relieve myself, carrying and cleaning the bedpan, and perhaps most distasteful of all to her, holding cigarettes up to my mouth so I could smoke—in her and Dad's bedroom, no less.

Dad had to go to work at the grain elevator, but each day he helped Mom get me out of bed, holding on to me as I stood up, gripping me tighter as I began to shake. I tried to take a few steps every day. I couldn't wait till I could walk again—better yet, to run again.

At night, Dad would sit by my side and read the **Salina Journal** to me. I still couldn't hold the newspaper for myself. Occasionally, our good friends Chet and Ruth Dawson came over to play bridge with Mom, Dad, and Norma Jean. They'd set the card table up as close to the bedroom as possible so I could see and hear, and be a part of the game. Other people wanted to come visit, but after the first few callers, I realized two things: First, Mom was working herself ragged, always trying to bake cookies, cakes, or pies for everyone stopping by. On top of the heavy workload she'd inherited with me alone, the extra effort was taxing her strength and energy. Second, I wasn't handling visitors too well myself. I appreciated their stopping by, but I just wasn't ready for them yet. I wasn't ready to see them, or for them to see me, and I certainly wasn't ready to talk about what had happened out there on Hill 913. It was a natural question for any visitor to ask: "Well, Bob, what in the world happened out there?" But I wasn't up for giving a play-by-play analysis. I didn't want to be rude to my friends, but after a few days of constant talking, I was exhausted, and I asked Mom and Dad to close the door and keep it closed. I didn't want to see anyone for a while.

I liked the Dawsons. Chet and Ruth had known me since I was a little boy; I'd worked at their drugstore from the time I was thirteen till I graduated from high school. But anyone else . . . I just couldn't handle properly and I was probably offending some.

Bringing me into a home environment placed an enormous strain on everybody in the family. Everything that the nurses, attendants, and the Red Cross Gray Ladies had done in the hospital still had to be done at home, but now there was just Mom and Dad and Norma Jean to help. Mom's sisters came by and helped with the housework occasionally, and other friends sent us all kinds of good food. Norma Jean had a job now, too, but she did as much for me as she could. She'd sit with me and feed me. But for the most part, after the initial excitement of my being home wore off, the day-to-day burden of caring for me fell mostly on Mom's shoulders.

It was tough on both of us. I soon got in the habit of calling out to Mom for every little thing. I hated to bother her, and sometimes I'd try to go as long as I could without interrupting her, but inevitably, every hour of the day, I'd need her for one reason or another.

From the family's standpoint, this was a new experience. Our family was accustomed to going, going, going. Mom and Dad didn't raise us to sit around and do nothing. We were a family constantly involved in everything—sports, community events, work, or something. Now, somebody had to be home near me, twenty-four hours a day. It was like having a baby at home—an adult baby, without the sense of wonder or the joy of discovery that comes with a child's first steps or first words. Rather than discovering a new world, I found myself repeatedly returning to the world of my past, languishing over what I'd lost, becoming angry over what I couldn't do, struggling to come to grips with the idea that I was not "whole" anymore.

On good days, when I'd accomplished some small achievement like taking an extra step or getting my finger to move, I was probably much easier to be around than on those days when nothing worked, when I felt like a colossal failure, a problem to my parents, a pain in the behind to everyone in the family. Sometimes, I'd simply lie in bed for hours on end and stare at the ceiling, determined to get up and walk, but discouraged by so little progress.

Worse yet, when I had a bad day I tended to bring everyone else down, too, especially my mom. Although she rarely said anything negative, I later learned that she'd sometimes vent to one of her sisters, or to Norma Jean or Dad. At her worst moments, she slipped into the kitchen, where I couldn't hear her crying, and confided through her tears to them, "Sometimes, I'm afraid we brought Bob home just to die."

Maybe that's why, as much as we loved each other, and as much as I was thrilled to be home and the folks were happy to have me there, it wasn't completely heartrending when the day came for me to go back to Winter General.

I had barely settled in at Winter General again when the word came that the army was transferring me on November 10, 1945, to Percy Jones Army Hospital, in Battle Creek, Michigan. By then, the army medical center at Percy Jones Army Hospital specialized in orthopedics, neurosurgery, physical therapy, and X-ray therapy. It was also the army's main center for paraplegics and amputees. Established in the summer of 1942, by the time I arrived there in 1945, the hospital housed more than eleven thousand wounded soldiers—twice the entire population of Russell, Kansas.

But Percy Jones Army Hospital had also established itself as a place where miracles could happen, where wounded soldiers got their lives—and often their limbs—back, even if it was through the modern science of prosthetics.

Maybe there was a miracle waiting for me in Michigan.

CHAPTER 17

Wishing for a Miracle

Tender, loving care was the hallmark of Percy Jones Army Hospital from the day it opened in 1942 to the day it closed in 1953, having served our country well. To most of the doctors, nurses, and orderlies at PJAH, every patient was special. Of course, in 1945, near the end of the war, the hospital staff didn't have much time to make that apparent. More than a hundred new wounded patients arrived by plane every day. Dozens more were off-loaded from Grand Trunk and Michigan Central Railroad trains into army ambulances waiting at the specially constructed boardwalk platform to transfer the wounded to Percy Jones.

PJAH was originally planned to house 1,500 patients, but by the time I arrived in November 1945, the population of wounded soldiers at the hospital and its facilities at nearby Fort Custer had swelled to ten times that number, peaking at 11,427. As the army's premier facility for neurosurgery, amputations, deep X-ray therapy, and the forming and fitting of plastic artificial eyes, Percy Jones performed more than 700 operations in one month alone. The hospital was also well known for its rehabilitative work in assisting disabled soldiers in reclaiming their lives. I wouldn't have guessed, as I was being prepared for another transfer, that Percy Jones Army Hospital would play such a central role in saving my life.

Ironically, the Battle Creek, Michigan, hospital opened shortly after the end of the Civil War as the "Western Health Reform Institute," founded on the nutritionally healthy living principles espoused by the Seventh Day Adventist Church. It was intended to be a "Temple of Health," to improve the health and well-being of individuals, both those who were sick and those who simply wanted to live more healthy lifestyles. It soon became known as the Battle Creek Sanitarium, and in 1876, Dr. John Harvey Kellogg took over as its superintendent. Kellogg's many innovations at the sanitarium included the use of radiation to help treat cancer patients; and, oh, yes, he invented a special breakfast treat for his patients, something he called "corn flakes." Dr. Kellogg's brother, Will Keith (W.K.), worked at the sanitarium for twenty-six years as bookkeeper and business manager, before leaving to establish the Battle Creek Toasted Corn Flakes Company, later simply known as Kellogg's.

The Kellogg brothers combined concepts of proper diet and exercise with hydrotherapy and other fashionable spa treatments, and soon attracted affluent guests from around the world to their facility. When fire destroyed the original wooden buildings in 1902, Dr. Kellogg designed an ornate replacement, a six-story Italian Renaissance–type structure, complete with enormous colonnade pillars across the front entry. The sanitarium reopened for business Memorial Day weekend in 1903. An opulent additional fifteen-story "towers" section designed by M. J. Morehouse of Chicago, was constructed in 1928, establishing the building as Battle Creek's first skyscraper. With a service staff of 1,800, the luxurious sanitarium hosted such guests as President Howard Taft, auto magnate Henry Ford, and many others seeking a restful retreat or hoping to discover the elixirs of life.

Following the stock market crash of 1929, the

sanitarium plunged into bankruptcy. It barely remained open, in receivership, during the Depression, and went on the public auction block in 1942. It was purchased by the U.S. Army for $2.5 million, an astronomical amount of money in those days, but probably a steal considering the enormous, elaborate structure's quality and beauty.

The army immediately converted the sanitarium into Percy Jones Army Hospital, to help treat the rapidly increasing number of war casualties. The hospital was named after Percy Lancelot Jones, an army surgeon who served in the Spanish-American War, and later pioneered modern battlefield ambulance evacuation operations during World War I.

Of course, I knew none of this when I landed at chilly Kellogg Field in a C-47 military transport plane, wondering what was ahead for me. I lay on my litter, as six or seven uniformed soldiers maneuvered me off the aircraft and into an ambulance for the short trip to the hospital. All I knew was what I'd heard from other patients at Winter General: I'd be in good hands at Percy Jones.

The stately entrance and the gorgeous tree-lined circular driveway out in front of Percy

Jones Army Hospital evoked a sense of quiet strength, tranquility, and serenity that belied what went on inside the building. Looking at the building as we drove up, if I hadn't known better, I'd have thought we were checking in to a luxury hotel. One could hardly imagine the hellish nightmares that had brought the soldiers inside to this place. More than 78,000 war-torn GIs were treated within the various buildings of Percy Jones Army Hospital during and after World War II, and 16,500 more vets were treated there during and after the Korean War. Each man went through those doorways carrying a lot of emotional baggage—and a little less of himself—than when he'd gone off to war.

I arrived at Percy Jones in my sealed cast, with my right hand strapped to a device that vaguely resembled a tennis racket and was intended to separate my fingers.

Inside the hospital, the brightly lit hallways and sparkling clean floors created a pristine yet upbeat atmosphere. I was placed in an orthopedic ward on the tenth floor with about a dozen other men, several of whom were in traction. The doctors and nurses greeted me kindly, though matter-of-factly. I looked around the ward and wondered, **With all these guys in**

here to care for, it will be a miracle if the doctor or nurse ever gets around to see all of us.

But Percy Jones was now known as a place where miracles could happen, and the medical staff was amazing. They called me by name, looked me in the eye when they talked to me, and did everything possible to remind me that I was important to them, that I, like every other soldier on the ward, was a person of dignity. The staff helped me in physical therapy, working with me for hours trying to pry my fingers apart, trying to get them to move, getting me up, helping me to relearn how to walk normally.

One drawback to being at Percy Jones rather than at Winter General in Topeka was that Battle Creek was a long way from Russell, Kansas. Mom and Dad were now hundreds of miles away, and it was nearly impossible for Dad and Norma Jean to visit on weekends. When I'd first come home, Mom moved to the apartment across from Winter General in Topeka. For her to move to Michigan wasn't quite as simple a matter. I missed her smile, her constant attention, and, of course, her cakes, pies, and other goodies.

About the time I was transferred to Percy

Jones, the Red Cross workers and many of the local townspeople came in to help decorate the facility for Thanksgiving. They set up beautiful seasonal displays in the common areas, complete with stacked wheat and corn stalks, shucked corn, bright orange pumpkins, and cornucopias filled with every sort of vegetable imaginable. Images of Pilgrims thanking God for helping them survive that tough first year in the new world were common.

In retrospect, the subtle implication was obvious. The 11,000 or more men in that hospital were in a new world, too, but unlike our Pilgrim forefathers, we were in a new world not of our own choosing, in a place we would have given almost anything to avoid. But there we were at Percy Jones on Thanksgiving 1945, each of us having come through a lot, most of us thankful to be alive.

We had a fabulous Thanksgiving dinner, including turkey and stuffing, mashed potatoes, and cranberry sauce. It was a far cry from Mom's home cooking, and I had to eat my meal in bed, but it was great.

To a man, we probably could have gone around the hospital and given thanks for something, although I must admit, it was difficult to

fight off self-pity that day. I couldn't help thinking of the many previous Thanksgivings when
I'd pulled my chair up to Mom and Dad's table,
which was laden with food. I could eat as much
as I wanted, and look for dessert as well. Then
kick back and relax, or go out and play some
basketball or get a bunch of guys together for a
sandlot football game later in the afternoon.
Now, I could barely digest my food, and I wasn't
going anywhere.

If Thanksgiving was proving to be that
wrenching, I certainly wasn't looking forward to
Christmas. Sure enough, as soon as the corn
stalks came down, the Christmas trees and other
decorations began going up. Throughout the
first few weeks of December, with the snow
falling outside, I could often hear the sounds of
Christmas carols resonating through the normally quiet hospital halls. One group of carolers
was particularly popular with the single fellows
at Percy Jones: the Masonic Service Center
Girls. The attractive young women came often
to serenade the patients.

I enjoyed their visits, and as a twenty-two-
year-old soldier, I still knew a pretty young
woman when I saw one. But that was the problem. Seeing those pretty young women struck a

different sort of fear into my heart over what I had lost. Throughout my high school and college years, and before being wounded in Italy, I hadn't exactly been a "ladies' man." Now I wondered if any woman would ever want to be seen out with **me**.

Nevertheless, I was glad to be in a place where progress seemed possible. A Colonel Mayfield noted in my record at the time, "It is obvious that this man has combined upper and lower motor neuron lesions, as well as deep injuries of the cervical spine. Treatment is to consist of PT [physical therapy], and passive and active exercises." At this point, there really wasn't much else they could do.

I dutifully worked with the physical therapists every day. I was beginning to make some progress, getting a bit more feeling back in part of my left hand, and taking a few small steps.

Then I suffered a major setback. Early on the morning of December 21, 1945, I awakened feeling soreness in the left side of my chest. At first, I thought that perhaps I had simply slept awkwardly—as it was hard to find a "comfortable" sleeping position in my condition. But then I realized that I was having trouble breathing, too. Suddenly, I felt pain in my chest cavity. I cried out, "Help!"

One of the guys next to me heard my cry and called for help. The doctors quickly checked me over and then rushed me off for X-rays. They discovered that I had developed a pulmonary infarct, a blood clot in my lung, precariously formed in such a position that it was affecting the pulmonary artery. If the clot broke loose and headed in the wrong direction—toward my heart—it would be deadly.

Treatment was going to be tough. It was a good bet that the blood clot had developed in the first place due to my months of inactivity, lying immobile for so long, consequently creating the dangerous conditions for clotting. But I didn't dare move until the clot was dissolved, for fear it would break loose and possibly make a beeline for my heart. The only option the doctors considered viable was to confine me strictly to bed and treat me with dicumarol, a strong blood thinner. Even at that, the doctors gave me only a fifty-fifty chance of survival.

For the next six days, the doctors injected dicumarol into my system. My temperature shot straight up, and I became extremely weak and light-headed. I wasn't sure where I was or what day it was. I was too sick to know or care. Even the optimists at Percy Jones didn't give me much hope.

On December 27, 1945, Second Lieutenant Robert E. Reker wrote to my dad to alert the family concerning the seriousness of my condition:

> **Dear Mr. Dole:**
> **We regret to inform you that your son, Robert J. Dole, who was admitted to this hospital on [November] 10, 1945, is seriously ill with Pulmonary infarction. At the present time it would appear that his recovery is somewhat questionable.**
> **If it is your desire to visit him at this hospital, you may feel free to do so at any time. You may, however, rest assured that everything has been and is being done to bring about his speedy recovery.**
> **Very truly yours,**
> **Robert E. Reker**
> **2d Lt, MAC**
> **Adjutant**

Somehow or other, I don't think Lieutenant Reker's letter was too comforting to my parents. It did, however, cause my dad to board a train for Battle Creek. Meanwhile, the dicumarol treatments continued, as did my fever.

There were no seats to be had aboard the train, which was jam-packed with servicemen on their way home from the war and others returning after a Christmas leave. My dad stood on his feet, hanging on to a handrail or ceiling loop, almost the entire way from Russell to Battle Creek. When he arrived, his feet were swollen and he could hardly walk . . . but he'd made it.

I was close to fading away on several occasions, but I kept snapping back. I was fighting hard, trying not to let go of life. I awakened every so often, then slipped back into darkness. Once, when I roused, I saw my dad standing at my bedside. It was the best Christmas present I could have received.

CHAPTER 18

You'll Never Walk Alone

The blood-thinning process lasted for weeks, continuing from December throughout January and into February. Several times each day, the nurses came by to draw blood to monitor my prothrombin level—the ratio between clotting in my blood and the thinning effects the dicumarol was having on my system. All the while, I lay on my back, weak, sick, and with what little strength I possessed dwindling away. It was the new year now; the war was finally over, and while many patients at Percy Jones Army Hospital had celebrated with fresh resolutions about what they intended to do in 1946, my future seemed in doubt.

With my condition growing worse rather than better, on February 12 the doctors decided to take me off the dicumarol. Almost immediately, my body was racked with pain again, especially in the lung area, where the blood clot had been detected. Worse yet, the fever came roaring back—with a vengeance. The next day, so did the chills, and more intense pain. Adding to the problems, I started coughing uncontrollably—deep heavy coughs emanating from the bellows of my lungs. The doctors put me back on the blood thinner on Valentine's Day, along with massive doses of penicillin, but my temperature continued to soar, this time to 106 degrees.

For the second time since surviving the war in Italy, it looked like I might not survive at home. I was coughing and sputtering, with each nagging cough ripping my lungs as though with a jagged-edged knife. The doctors were concerned that if the blood clot didn't kill me, pneumonia would.

Early in March they ran out of ideas. The hospital had already called in my family members, and it seemed certain they had made the trip from Russell to Percy Jones to stand by my bed and watch me die. Even their being there was an exercise in futility; I was too feverish to know that anyone was in the room, not even my own family members.

Yet in a way, the fact that they had made the trip proved providential.

The doctors held out little hope. There was, however, one extremely remote chance. At Rutgers University, Dr. Selman Waksman, a biochemist, and his research assistant, Albert Schatz, had been experimenting with a new antibiotic in their quest to overcome penicillin-resistant bacteria. As a result, they had found a drug thought to be useful in treating tuberculosis. The new drug was known to be powerful, but it had not yet been tested on enough human beings to predict its effectiveness with any accuracy. Of the known side effects, some people had experienced problems with their kidneys, others had lost their hearing, at least one person had gone blind, one person had died, and others had survived, but at the cost of perpetual dizziness and nerve damage. The drug was known as streptomycin. Dr. Frank Solomon, an outstanding vascular physician, thought the potential benefits of the drug outweighed the risks—especially in my case.

The army had received a small amount of streptomycin for experimental purposes at Percy Jones, but they refused to administer it without the written consent of the next of kin to authorize the treatment. The army said something such as, "Take some time to think about it, Mr. and Mrs. Dole. Keep in mind that we cannot assure you that this drug will work. Your son may not respond well to the treatment. Furthermore, we can't guarantee that he will not incur some physical damage, or that he will be able to walk, or talk, or do anything else after receiving the streptomycin. We really don't know for certain how your son's body will react to this drug. But we believe that since the penicillin has not worked, this may be our only hope."

Mom and Dad signed the forms authorizing the hospital to administer the drug to me, and basically use me as a human guinea pig. There really were no other alternatives. To do nothing would have resulted in my certain death. That's why it was providential that Mom and Dad made that trip to Battle Creek. Had they not been there, the hospital might not have been willing to try the streptomycin.

In 1952, Dr. Waksman was awarded the Nobel Prize for his work in physiology which had led to the discovery of several antibiotics, including streptomycin. A few years later, the "wonder drug" streptomycin was in wide use in the battle against tuberculosis and other penicillin-resistant bacterias. A few decades later, the strong drug was pulled off the market after it was blamed for causing kidney problems and deafness in some patients.

On March 6, 1946, four days after the doctors injected the streptomycin into my system, I was able to sit up in bed. Soon after that, the doctors permitted me to sit in a chair for short periods of time. My brother, Kenny, said later that when I first responded to the medicine, I woke up and sent him out to get me a chocolate milkshake. Could be, but I have no recollection of that. All I know is that on March 6, I turned a corner on my road to recovery. It was still going to be a long, difficult journey, but from that day on, I knew I was going to live. Apparently, I still had more work to do, a greater purpose to fulfill.

There were no shortcuts to getting back on my feet. I had to start the rehabilitation process all over again, waiting while the nurses and orderlies helped me out of the traction, wiggling my toes to regain some feeling in them, easing myself off the bed, touching the floor with my feet, and learning to stand on my own two legs once more. Each part of that process took long hours, sometimes days to accomplish. My right arm was still in a "soft" cast, wrapped in strong, stiff material, and propped up to about a forty-five-degree angle by a triangular piece that looked similar to the rack used to align billiard balls. Each finger was separated by a sling-type wrap, which was meant to keep my fingers extended so they wouldn't curl up into a gnarled claw, as they tended to do.

Occasionally, when the nurses took the coverings off my right arm, I could see that it was pretty badly banged up and disfigured, with the bones jutting awkwardly. The nurses massaged my right arm, trying to help me get some feeling back into it. The doctors expressed concern that ankylosis might set in—that my arm might "freeze" in one position—due to the lack of movement. My elbow had already become fused because I'd been in a cast so long. My left arm remained next to useless due to the spinal injury affecting the nerves, so it was impossible for me to use crutches or a walker. I had to learn to balance myself without using my hands or arms.

I took slow, baby steps at first, shuffling along the floor from one end of my bed to the other. Sometimes I'd get adventurous and try to walk across the room. One day, I was attempting to make it back to my bed when I lost my balance and fell. It was one of those awful moments when the whole place suddenly fell silent. No doubt everyone was wondering if I had dropped dead. Then several guys dressed in pajamas came running to help, as did the nurses and orderlies.

I shook my head. "No, I'm okay," I said. "I need to get up by myself." I didn't want to cause a scene or be a spectacle, but I didn't want to depend on other people forever, either. I had to start learning how to do things for myself. I lay there on the floor in my pajamas, gathering my strength and my will, and then, inch by inch, struggled to my knees. I rested a while, then teetered and tottered until I could finally stand up. The guys who were standing around watching burst into applause. Everyone at Percy Jones was fighting to overcome something; we were all rooting for each other. When one of us accomplished a feat, no matter how small, we all shared in the elation.

My spirits brightened with each new success. Slowly, as my health returned, so did my sense of humor. Some people may have thought that I

was trying to conceal or cover my pain and inse-
curities with humor, hiding behind a deadpan
delivery of one-liners to keep from dealing with
my incapacitation. I don't think so. Humor had
always been a part of my normal repertoire. Like
my dad, I had always had a wisecrack, a quip, or
a funny line to share, and I shared them now
with anyone who came by my bed to visit. The
nurses smiled politely when I first began joking
with them; they were accustomed to soldiers'
talk, and they didn't let it keep them from doing
their jobs.

One army nurse particularly stands out in my
mind. She had been in a car wreck, and the force
of the crash had slammed her into the car's console, allowing the radio knob to punch a hole in
her forehead. Technically, she was a patient at
the hospital, but during her long recovery period
she also volunteered to help others. She came by
the ward frequently, and often helped feed me
my meals. Her selfless service was a shining example that a person doesn't have to be perfectly
whole to be of service to someone else. All that
is needed are care and compassion.

The nurses got me out of bed each day and
allowed me to sit in a wheelchair. Because my

arms and fingers refused to function, I was unable to roll the chair myself, so I had to rely on the nurses to wheel me around the ward; other patients helped, too. Funny how having to rely on someone else changes a person's perception of himself or herself—I soon acquired a whole repertoire of jokes regarding my inability to move in the chair. Eventually, the nurses warmed to my dry, droll, low-key sense of humor. They even started wheeling me around to other wards in the hospital to encourage other guys who had been wounded.

Visiting other wounded soldiers was good therapy for me, too. I'd crack a joke or poke fun at some situation or physical limitation that we shared. And we shared a lot. Everywhere I went around the hospital, I found other soldiers whose dreams had been literally blown apart by the war. Everyone had a story; in nearly every ward, in every bed, there was a person who hurt just as badly as I did—and many who had suffered far worse than I had.

Still, when I rolled around their wards, the wounded knew that I wasn't just some guy who was trying to conjure up false sympathy for them; I could genuinely empathize with them because I had experienced similar pain. Interestingly, my suffering provided me the credibility to encourage others in getting through theirs. Apparently, I had earned something of a reputation, too, through the streptomycin experiment. Several of the nurses and doctors jokingly referred to me as "a soldier who refused to die when he was supposed to."

One fellow whom I stopped to visit regularly, trying to cheer him up, was Joe Brennan, a soldier from Chicago. Joe was paralyzed from the waist down. The poor guy had large, open bedsores from lack of movement. He'd lie in his bed smoking a cigarette, and we'd talk and laugh together. Being a big-city boy, Joe loved to tease me about being from Russell, Kansas. "Do you people get sunshine there yet?" he'd ask.

I'd give it right back to him. "Yeah, we do. At least we can see it when it shines."

Joe had a great sense of humor, even about his disability. The other guys often gave him a hard time. "Hey, Joe. How about running down to the cigarette machine and getting me a pack of Lucky Strikes?"

"Yeah, I better do the running. If you try it, they'll pull your disability and call you back up to active duty."

Unfortunately, Joe didn't have to worry about

returning to active duty; another American hero, he died within a few years after the war. Joe always had a smile on his face. His grit, bravery, courage, and sense of humor made for a good example of someone who kept a positive outlook despite all he went through.

I awoke every morning at Percy Jones, hoping that this would be the day for my miracle. But I refused to wait around for it to happen. Each day, I worked as long and as hard on my recovery as I could. I went to physical therapy, did exercises to strengthen my legs, and worked constantly to attempt to get the fingers on my right hand to move. When they'd stubbornly remain in place, I'd work on my left arm, which although not functioning well, had a bit of feeling at least.

The army physicians who worked most closely with me refused to give up. They tried everything to get my right hand to move. The nurses gave me repeated muscle-relaxing baths and hot wax treatments, in which they'd coat my arms and hands with the wax in hopes that the heat would cause my fingers to loosen enough that they might move. Nothing seemed to work. But we kept trying.

The drudgery of the daily grind in the hospital was often relieved temporarily by the appearance of some famous celebrity stopping by to visit and entertain the troops. It was a time in American history when big-name Hollywood stars and chart-topping "Top Ten," or "Hit Parade" musical artists and others in the entertainment community recognized and honored the soldiers who had paid such a high price to protect the world's freedom. Busy celebrities such as radio (and later television) star Jack Benny, actor Alan Ladd, and world-famous drummer Gene Krupa generously gave of their time, wending their way through the hospital wards, talking with the soldiers, and signing autographs.

Dinah Shore would sit on the edge of a soldier's bed, talk to him about his hometown and his family, then sing one of his favorite songs for him. Gene Autry, known as "the Singing Cowboy," made numerous visits to Percy Jones. A Who's Who of the music business came to the hospital, including Rosemary Clooney, the Mills Brothers, Guy Lombardo, Duke Ellington,

Eddie Cantor, Stan Kenton, and many others. Some of the artists performed in concerts held in the Percy Jones gymnasium—it was there that the large troupes such as the Ted Mack Amateur Hour performed live radio broadcasts—but many artists performed in the wards as well, taking their music and messages of hope to soldiers who couldn't come to them.

Unquestionably, one of the most memorable celebrities to visit Percy Jones Army Hospital was Bob Hope. The entertainer worked nonstop, visiting with as many of the patients as possible, personally going into the wards to greet the guys who were unable to get up and go to the show. Then Hope and his group of traveling troubadours and his bevy of beautiful young women put on a show for more than an hour and a half, which was met enthusiastically with applause and shrill whistles from the guys.

About that time, I decided to do some whistling myself. I noticed how some of the guys could create great sounds just using their tongue, teeth, and cheeks. I had started whistling tunes back in high school, sometimes while I was cleaning up after hours at Dawson's Drugstore. But at Percy Jones, I had plenty of time to practice.

I whistled constantly, too, probably driving

everyone around me half crazy. I loved whistling some of my favorite songs, such as "Deep Purple," which my sister taught me as a boy, "In the Mood," and other popular songs of the day. I loved anything by Frank Sinatra, and I was especially fond of a new song that I'd heard quite often since coming home from the war—a song called "You'll Never Walk Alone."

"You'll Never Walk Alone" became the signature song from Richard Rodgers and Oscar Hammerstein's 1945 Broadway musical play **Carousel**. In the show, a bombastic carnival barker, Billy Bigelow, marries the sweet, demure Julie Jordan. When the young couple discovers that they are expecting a baby, Billy foolishly attempts a robbery to get some money. In the process, he is killed.

At the "Pearly Gates," Billy is given one more chance—if he can prove his love for Julie and the baby, he'll find heaven's gates flung open wide. The song "You'll Never Walk Alone" is sung at a pivotal point in the story, and then again at Billy's daughter's high school graduation at the end of the musical, as the powerful and inspirational finale.

From **Carousel**'s first performance on Broadway in 1945—when the knowledge was just starting to sink into the national psyche that many dads, sons, and brothers would not be coming home—"You'll Never Walk Alone" took on a meaning far beyond the context of the musical. Many people who had loved ones far away from home found solace in the simple yet profound message:

> **When you walk through a storm, hold**
>   **your head up high,**
> **And don't be afraid of the dark.**
> **At the end of the storm is a golden**
>   **sky,**
> **And the sweet silver song of a lark.**
> **Walk on through the wind, walk on**
>   **through the rain,**
> **Though your dreams be tossed and**
>   **blown,**
> **Walk on, walk on with hope in your**
>   **heart**
> **And you'll never walk alone,**
> **You'll never walk alone.**

Over the years, the song has been recorded by dozens of musical artists of every style—from

Chet Atkins to Louis Armstrong to Elvis Presley to Pink Floyd—but the rendition that moved me most deeply and spoke to my soul was the one done by Frank Sinatra.

I'd listen to Sinatra singing "You'll Never Walk Alone" and it would remind me of my own delayed dreams. If there was anything I wanted to do, it was to walk on, to walk normally, to be able to run—I'd always loved running, even as a young boy—to play basketball again, to be able to walk down Main Street in my hometown of Russell, to stop by and see my buddies Chet and Bub Dawson. "Hey, Chet, how 'bout setting me up a big, thick chocolate milkshake."

It was all a dream.

But the message of the song reminded me that even though my dreams had been tossed and blown, if I walked on, if I kept struggling to make my legs move, to put one foot in front of the other one, I could do anything. I could walk, and all of heaven would help me; my upbringing would help, my grandparents, schoolteachers, athletic team coaches, university professors, my boot camp sergeants, my buddies who went out that morning near Castel d'Aiano, up Hill 913 with me, the doctors and nurses in the hospitals, the thousands of other men and women who had laid it all on the line to win that war—they'd all be there, cheering me on. No, I'd never walk alone, but I could **walk**; I **would** walk; I **had** to walk, if not just for myself, then for all of them, too.

CHAPTER 19

Friends Who Last

In one sense, the pessimists had been right: I'd never walk again as I had before. But in another sense, they were dead wrong. I would walk, but I'd never again walk alone. They would always be there, all of them—my family, friends, and fallen heroes—each step of the way.

More determined than ever, I began to see progress in my steps. Slowly—agonizingly so—I attempted to regain the use of my hands, as well. I started by lifting a paper with my left hand, then each day, lifting a little more weight. Cups and glasses remained difficult, and I hate even to think of how much food and drink I spilled on myself in the hospital. It had been a full year since the last time I'd raised a fork to my mouth, but I was getting there with my left hand. The tough part was that I had feeling only in the last two fingers. If you want to get an idea what a difference that makes, try buttoning a shirt using only those two fingers, or tying a shoelace, or a necktie.

Later on, I experimented with all sorts of articles of clothing that I wouldn't need my fingers to manipulate. I tried every sort of snap and zipper available at the time; later I'd try jackets that closed with Velcro, and slip-on shoes. I tried clip-on ties, but they always looked like a weird appendage attached to my shirt. I discovered that cuff links work better than buttons on my shirtsleeves. Over the years, eventually, I even learned to tie a necktie, though I rarely loosened my tie in public, for fear I'd never get the thing straightened again.

The right hand was still rebelling, each finger operating as though it had a mind of its own. I got to the place where I simply regarded it as a "helping hand," sort of as a backstop against which my left hand could work and actually do things. That wasn't much, but at least I still had the second hand. Many men in the amputee wards at Percy Jones would have been grateful even for that.

My walk was improving, too, so much so that I now ventured down to the second floor on my own. I moved slowly, and stayed to one side of the hallways, to avoid having to get out of someone's way in a hurry. It took a while, but I eventually made it to the second floor. That's where the real action was—at the bridge games that went on almost perpetually from early each morning until lights-out each night. Occasionally, I'd even join in on a game, although someone else would have to shuffle and deal the cards for me. Mostly, I just watched and enjoyed the camaraderie.

One of the best bridge players in the entire hospital was a young Japanese-American with only one arm. His name was Danny. Daniel Inouye.

On April 21, 1945, one week to the day after I was wounded in Italy, Daniel Inouye led an assault on several machine gun emplacements on a steep hill called Colle Musatello, an important junction near San Terenzo, Italy, a short distance from Castel d'Aiano and Hill 913, the place where I had been hit. Leading his platoon up the hill, Danny's right arm was shattered during the attack by a blast from an enemy rifle grenade. He also took a bullet in his leg, one in his side, and another that came out his back.

The fact that Inouye was fighting in Italy at

all held a certain irony of its own. Of Japanese ancestry, born and raised in Hawaii, Danny Inouye was a seventeen-year-old college student when he witnessed the dozens of Rising Sun aircraft droning over his parents' home on their way to bomb Pearl Harbor on Sunday, December 7, 1941.

A proud American patriot who had been enrolled in a Red Cross first-aid training program, Danny rushed to Pearl Harbor to offer his assistance. He jumped right into the death and destruction and began to help treat the hundreds of wounded. Soon after his courageous, sacrificial service to our country, the United States government declared Inouye and all Japanese-American men "enemy aliens." Many Japanese-Americans were rounded up and incarcerated in U.S. prison camps. Insulted, Inouye petitioned for the right to enlist in the army to fight for his country. Eventually, the army conceded, and created an all–Japanese American unit known as the 442nd Regimental Combat Team. The members of the 442 fought tenaciously, as though they had something to prove. Indeed, they did. By the end of the war, the 442nd had become one of the most highly decorated units in U.S. Army history.

As Daniel and I compared notes and commiserated in Percy Jones Army Hospital, we developed a close friendship. Besides the coincidence of our being wounded within days and miles of each other at the close of the war, we had other things in common, as well. I'd lost the use of my right arm, and what was left of Danny's right arm had to be amputated. We were both learning how to do everything from smoking a cigarette to opening a letter with one hand.

Slight of build at the time, Danny weighed a mere ninety-three pounds when he entered Percy Jones. He had a tremendously positive attitude, and he was the best bridge player I'd ever seen. I marveled as I watched him whip everyone in the ward, playing with only one hand, his cards placed in a rack in front of him. We'd laugh a lot after a bridge game as he'd try to explain to me that it was skill not luck that had brought him the victory.

Years later, Daniel Inouye would even give me indirect credit for getting him started in politics, although I've never made such a claim myself. The way Inouye tells the story, he and I were sitting around playing cards one day, speculating about our futures . . . or lack of them.

"What are you going to do with yourself?" Danny asked.

Supposedly, I replied, "When I get out of here, I'm going to go to law school, become an attorney, run for the state legislature, and when the opportunity comes along, I'm going to run for the U.S. Congress."

Danny says that inspired him to do the same. Whether Danny's story is accurate or not, I'm honored that he'd credit me with playing such a prominent role in his life. Years later, when commenting on his successful career in the U.S. Senate, Daniel said, "I followed the Dole plan, and I beat him!"

Maybe so. Hawaii became our nation's fiftieth state in 1959, and Daniel Inouye became the new kid in Congress that year, as one of the state's first members in the U.S. House of Representatives. I didn't show up in Washington until the following year, as a congressman from Kansas. In 1962, Daniel Inouye won one of Hawaii's senatorial seats, where he has served for more than forty consecutive years. It was a pleasure serving with Danny in the Senate. One of the last letters I wrote on my final day in the Senate, in June 1996, was to Danny, jokingly appointing him as chair of the "Percy Jones Alumni Caucus."

Another outstanding person I met at Percy Jones was a young soldier named Phil Hart. Phil was one of the best men I ever knew.

He had served in the army since 1941, and had been seriously wounded during the D-day assault on Utah Beach in Normandy. He had already been in the hospital for some time when I arrived. Because his condition was on the upswing, Phil was more mobile than most other patients. Consequently, he was constantly trying to serve his fellow soldiers, running errands for guys, lighting cigarettes, or procuring extra snacks from the dining room.

Phil was married to Jane Briggs, daughter of Walter O. Briggs, the former multimillionaire owner of the Detroit Tigers Major League Baseball team, and Phil had no qualms about getting soldiers' visiting family members tickets to Tigers games.

He later served as lieutenant governor of Michigan and went on to be elected as a Democrat to the U.S. Senate in 1958. He served in

Congress for twenty-two years, and in 1971 was the only member of that august body to sport a beard. His wife, Janie, was a pilot, and one of the first licensed female helicopter pilots in Michigan. Beyond that, she later became one of the "Mercury 13," a group of American women who passed the rigorous tests to become astronauts, but never made it into space.

Phil and I were diametrically opposed politically. He was the quintessential liberal, and I was just the opposite. Yet he and I remained good friends. He was one of the kindest, most compassionate people I've ever known, a politician who proved that politics could indeed be an honorable profession rather than a cutthroat career.

When Phil Hart died in 1976, the nation mourned the loss of the man known as "the conscience of the Senate." In 1987, the Senate Office Building in which I maintained my Washington office was renamed the Phil Hart Senate Office Building. Ted Kennedy once remarked about Phil Hart that he was the finest senator ever to serve our country. I certainly agreed.

Looking back, I find it amazing that the three of us—Phil Hart, Daniel Inouye, and I—three wounded soldiers who became such good friends through our common suffering, would all one day serve our country as United States senators. Beyond that, we all remained close friends. Something about our World War II experiences and our time together at Percy Jones Hospital created a bond among us that no partisan politics could ever separate. Throughout our lives, each of us always felt a debt of gratitude to the men and women who served at Percy Jones, and to those soldiers who arrived there with so little hope but who were able to leave with a new lease on life. In fact, in 2003, the Percy Jones Army Hospital (now a federal office complex) was renamed the Hart-Dole-Inouye Federal Center.

Of course, in 1946, I'm really not sure that I ever dreamed of serving in the Senate, much less having buildings named after me. In fact, at the time, I had grave doubts that any of my grand dreams would ever come to pass. My greatest fear was that I'd be languishing in a wheelchair, selling pencils on street corners, scrounging to support myself.

CHAPTER 20

The Longest Walk

Perhaps concerns about making a living prompted me to start looking for some ways to supplement my lieutenant's salary, while still a patient at Percy Jones. That, and the fact that during the war, America's major auto companies, such as Ford and General Motors, had transformed their auto-making plants into factories that produced war machinery. Such universal support for the war effort may be difficult for some people to understand nowadays. Imagine, if you can, the entire computer industry, for example—IBM, Apple, Microsoft, Dell, Hewlett Packard, everybody—retooling its production facilities during the war with Iraq in order to make only equipment used to further the war effort. That's what the auto industry and many other industries did during World War II.

The war affected nearly every facet of our society.

By 1946, however, with the war over, General Motors and the other auto companies began shifting back to their normal production plans. The American public was hungry for new automobiles, but before the new models started appearing in the showrooms, GM provided the first Buicks and Oldsmobiles off the assembly line to seriously wounded soldiers, and particularly to the amputees at Percy Jones Army Hospital. Each car was specially equipped to compensate for the individual disability of each veteran. Then the brand-new automobiles were delivered right to Percy Jones, ready for their new owners. The disabled soldiers still had to pay for the vehicles.

When I learned that an Oldsmobile dealership needed a salesman to pitch the cars to servicemen and women inside Percy Jones, the entrepreneurial side of me shifted into gear. I thought, **This sounds like a good deal for both of us.**

My job was to sell the cars to fellow patients, and I would make a 6 percent commission on every vehicle sold. Selling cars inside the hospital was easier than it might seem, since many of the patients had some savings. I had a captive audience. Moreover, most Americans hadn't seen a new model automobile since the United States entered the war, so many of us were excited about driving the new cars. I sold six cars, but when the army brass learned of it, they pulled the plug on my venture. Nevertheless, I later qualified for a car myself under a different program, in which the government paid $1,500 toward the purchase of one of the specially equipped vehicles. I bought a car from Ernie Drube, a Russell Chevrolet dealer and World War II veteran.

As hard as all the dedicated doctors, nurses, staff, and volunteers at Percy Jones tried, I just wasn't making much progress—at least not as quickly as I wanted to make it. I still had dreams of getting back to KU and playing basketball for Phog Allen, although I admit, the dream was fading. After being at Percy Jones for more than six months, I could barely walk, had no movement in my right arm and hand, and had very little use of my left arm and hand. The slow, monotonous physical and occupational therapy programs were simply not leading to the miraculous recovery I had hoped for.

When an opportunity came for me to have a thirty-day leave, starting on May 29, 1946, I went home. Once again, Mom and Dad dropped everything, and adapted their daily lifestyle to mine; they moved out of their own bedroom to the back room, and again installed a rented hospital bed for me in their bedroom. I was more mobile during this visit than I'd been during my first trip home, but I still had to be dressed by someone else and helped to the bathroom and all that. But unlike my first visit, I was now able to sit at the table for meals. Norma Jean helped feed me, since my left hand could not always be counted on to get the fork from my plate to my mouth with the food still on it. It really frustrated me when I couldn't hold a cup, or when I'd make a mess because I couldn't keep from shaking. Sometimes, I'd simply leave the table in dismay and go back to my room without eating.

Of course, I never went hungry. Mom waited on me hand and foot. We'd made it this far; she wasn't going to allow me to fail now. Before long, she'd bring food to the room and try to encourage me. "It's okay, Bob. You're doing well. It's just going to take some time."

Maybe so, but time was fleeting. I was approaching my twenty-third birthday, and still had to be cared for all the time. I became more determined than ever to develop my own rehabilitation program. It was my problem, and I was the only one who could do anything about it. The doctors had done their best; and while my family and friends were willing to do anything to help, it was now up to me.

My brother, Kenny, was back from the war, living at home, and with his help and Dad's, we developed a makeshift piece of exercise equipment, made from some pulleys and lead window sash weights. The contraption looked like two springs. On one end of the pulley, Dad attached a handle, similar to the kind a person might find at the end of a water ski rope. He fastened the other end of the pulley to the side of the garage, behind the house. I'd work out in the backyard for hours, trying to get my arms to function. Nothing else mattered. I had to get my arms working on my own. Often I'd work so long that Mom would come out and plead with me to come inside to eat something.

"I'll be right there," I'd say. Then I'd work for

another half hour to an hour on the pulley.
When I wasn't using the pulley, I carried a
palm-sized rubber ball with me all day long,
squeezing it constantly, trying to build up some
strength in my left hand.

Despite Mom's home cooking, my weight
was slow in coming back, still in the 120s. Worse
yet, I felt weak and frail. Nevertheless, I set
about trying to rebuild my physical strength and
stamina, hoping to recover the body I'd lost, or
at least some measure of it.

My good friend Adolph Reisig was one guy
who never gave up trying to help. He designed a
heavy, lead-weighted arm brace, lined with felt,
which I could wear on my right arm. I wore the
unusual device in hopes that it might straighten
the arm. While it didn't work, Adolph's efforts
meant more to me than anything else.

Not surprisingly, my rehabilitation increasingly took over our family's priorities. Any former schedule that we'd maintained as a family
was shucked, and we operated according to the
needs of the moment, especially my needs at
the moment. Chet and Ruth Dawson often
came over late at night, long after they'd closed
the drugstore, for a late-night bridge game.
Mom would often be doing a load of washing in

the midst of the game. Poor Dad. He still got up and went to work every morning, no matter how late we had stayed up the night before.

Mom and Dad had kept the old record player that I'd had at KU; it was in tip-top condition. One day, I came upon the phonograph, and pulled out some records, including my new favorite—Frank Sinatra's rendition of "You'll Never Walk Alone." Using my right hand to help me maneuver the disc, I awkwardly fumbled with the record, eventually working it onto the turntable. I did my best to let the tone arm and needle down onto the record without scraping it. A scratching sound, followed by the sound of the needle in the grooves told me that I'd succeeded in getting the record to play.

I sat back and listened to Frank.

Walk on, walk on, with hope in your heart . . .

The song played all the way through, but I wasn't satisfied. I wanted to hear it again—louder. I turned up the volume on the old phonograph and listened to the words of the song again.

When you walk through a storm,
Hold your head up high . . .

I played the song over and over and over. No doubt, Mom, Dad, Norma Jean, and Kenny got sick of hearing it, but they didn't complain. They noticed the way the song lifted my spirits—and that lifted theirs.

Walk on, walk on with hope in your
   heart,
And you'll never walk alone,
You'll never walk alone!

I was more determined than ever to walk. Oh, sure, I'd been taking baby steps for a couple of months now in the hospital, but that was more like a slow shuffle. I wanted to walk the way other people walked, the way I used to walk, the way I believed that I could walk again.

I started expanding my horizons a bit in the backyard, taking a few steps out by the garage. Then one day, I decided to venture a little farther, to the edge of the yard. Day after day, I stretched my limits, pressing just a little farther, doing a little more than I had done the day before. Sometimes people search high and low looking for an easy solution to the problems they face or the tough situations they encounter. I'm convinced that most of us would do well just by getting up today and attempting to do something that we didn't get to do yesterday. The moment you stop stretching, atrophy sets in, and the law of entropy (everything tends toward disintegration) takes over. No matter the storm, you've got to get up and walk on.

That's what I kept telling myself, anyhow. Both of my legs sometimes quivered as I walked—the doctors still had not found a way to stop the tremors from occurring, and the spasms often struck with no advance warning—but I forced myself to keep moving. I kept the words of "You'll Never Walk Alone" in mind, and kept the record player blaring half the night while I worked on trying to get my arm to move, squeezing the rubber ball, or simply lifting my legs off the floor.

Frequently, I'd whistle along with the song as the record played. I've never considered myself a good singer, but I'm sure there were times when maybe I hummed along with Frank, even if I didn't belt it out vocally. Almost before the song was done playing, I'd be back at the phonograph, calling out to nobody in particular, "Pretty good, eh? How 'bout some more music?" I never waited for anyone to respond. I just dropped the needle into its familiar groove once again.

Then one day, I decided I was going to do it. I was going to walk down the street. I'm not sure that I had a firm destination in mind, I just wanted to see if I could do it. I didn't tell Mom or Dad or anyone else. I just nudged the screen door open and stepped out onto the porch. It was a bright, beautiful Kansas springtime day, and the light caused me to squint at first. I felt a slight quiver in my legs, but I ignored it.

The end of the porch seemed a mile away, as I eyed it carefully, planning my trek as though I were setting out on a journey of a thousand miles—I guess in a way, I was. I stepped slowly over to the edge of the first of two steps down. This was going to be the toughest part, getting down those steps.

I awkwardly lifted my right leg down, placing it on the first step while my other leg was still on the porch level. Ever so slowly, I slid my left leg off the porch. Holding the leg stiffly, I swung it down to the step. There I was in limbo, not quite on the porch, yet not quite on the sidewalk

either. It was a precarious place to be, so I
quickly screwed up my courage and moved my
right leg to the sidewalk. I felt my leg shake
slightly, and for a moment I almost lost my balance. I quickly brought the other leg down to
the sidewalk level.

If I had done nothing else that day, I would
have already considered my efforts a success.
But now that I was on the sidewalk, there was a
whole world out in front of me . . . well, at least
a whole block of straight, flat sidewalk. I stepped
slowly and carefully, keeping my eyes constantly
on the ground as I walked toward the corner of
Maple and Main. One crack in the sidewalk or
one stone catching my shoe could be enough to
cause me to stumble, to trip, and to fall. I kept
my head down, my eyes riveted on the area directly in front of me. I felt as though other people might be watching out their windows or
from across the street. I thought for sure that I
could feel their eyes burning into me as I struggled down the street. It didn't matter. And no
doubt I made quite an interesting spectacle,
with my perfectly pressed double-pleated slacks,
a collarless shirt, and a short-sleeved sweater
draped over my back. My right arm lugged the
heavy brace Adolph had made for me. After

what seemed like an hour—though I'm sure it wasn't nearly that long—I reached the street corner, a pivotal point.

Was I going to turn around and go back home, basking in my successful effort? Or was I going to carry on and attempt to head up Main Street? I knew exactly where I wanted to go now. I straightened up, did a right face, and started up Main Street, toward Dawson's Drugstore.

Cars were parked in forty-five-degree-angled parking spaces on both sides of the street as I made my way up the sidewalk just off the red-brick cobblestone road. Several people called out to me, "Hey, Bob! How are you?" Others waved and said, "Good to see you!" I tried to make some sort of gesture, as a means of acknowledgment—they'd never understand that I couldn't raise either arm in the air to wave back at them, and I didn't want to offend anyone. So I nodded my head to let them know that I'd heard their words of encouragement and appreciated them.

Some people wanted to stop and talk, but I was on a mission. I kept moving forward. Others stopped and talked among themselves as I walked by. Sometimes they talked much too loudly. "Oh, that's Bob Dole? He looks awful, doesn't he? Yes, I heard that he got banged up pretty badly somewhere in Italy. Oh, he's so skinny. I wonder what happened to him."

The warmth of the sun felt good, but I was feeling a bit too warm. Moreover, I hadn't anticipated how dreadfully tired I'd feel. This little bit of walking, and I was as fatigued as I'd ever been after playing a complete high school football game.

But then I heard the words in my mind . . .

Walk on, walk on . . . and you'll never walk alone. . . .

I took another step, then another. Finally, I looked up from the sidewalk and saw the sign for Dawson's Drugstore. I'd made it.

I stepped up to the entrance, then realized I might have a problem opening the heavy front door. Fortunately, just then someone came out of the drugstore and I was able to catch the door with my foot. That's all it took. Somebody inside caught a glimpse of me and came running to swing the door wide open.

I stepped inside, stopped, looked around, and for a moment I was suddenly overwhelmed with emotion. The place looked almost identical to the way it looked the day I last set foot in it. The row of stools at the soda fountain counter was still there. The wooden-topped tables where the older fellows used to play checkers half the day, the booths in the back, the KU and K-State pennants, the slow-moving ceiling fans . . . it was almost like stepping back in time . . . and for a long moment or two I wished that I could go back, back before the war . . . when life was good, work was fun. I could see myself behind the counter, tossing the ice cream into the air, catching it just in the nick of time, putting on a little show for my friends . . .

"Do ya want the flip in it?"

"What?"

It was Chet Dawson. "Do you want the flip in your chocolate malted?" Chet said with a laugh. "Come on in, this one's on me." Chet put a hand on my shoulder—gently—and helped me move toward the ice cream and soda fountain counter.

Other people in the drugstore weren't sure how to respond to me at first. They seemed to hold back a bit, waiting, watching as I made my way to one of the stools and struggled to edge myself onto it. Chet and Bub Dawson helped me adjust my weight on the stool, and I pulled my foot up to rest on the foot rail that ran along the front of the counter. I was back.

"Hey, Bob!" someone called. "So good to see you."

"You're looking great, fella," somebody lied.

A few people came close enough to say hello that I didn't have to turn awkwardly on my stool to speak to them. It was wonderful to be able to see the faces of old friends.

And no chocolate milkshake ever tasted better than the one Chet made me that day.

Of course, there was sadness, too. The normal course of conversation inevitably involved talk about the war. And that meant talk about the boys who had arrived home safely . . . and those who hadn't. One of my best friends, Bud Smith, hadn't. A number of other Russell boys would never be coming home as well; most of them I knew personally, although there were a few guys I'd never heard of before the war. But they were all heroes to me now.

In an effort to be complimentary, no doubt, one of the folks at Dawson's said, "Bob, you're a real hero."

My response was almost a knee-jerk answer. I didn't mean it to sound curt or gruff. The people in Russell understood. It was the truth. "The heroes are still over there," I said.

Many of them were buried in the large World War II cemetery near Florence, with its row after row of white crosses lining the landscape. Others were buried near Normandy, in the Philippines, and throughout Europe, and of course, still more were buried at Arlington National Cemetery, outside Washington, D.C. To me, those guys were the heroes. I still believe that to this day.

I also became more aware of the awkward feeling people in public would have around me for the rest of my life when they first met me, or saw me in person after a long absence. Some people like to hug when they greet one another; in some countries, friends or family members kiss on both cheeks; in Russell, we usually just shook hands. A man's strong grip, a confident look, squarely in the eyes—people placed a lot of stock in those gestures.

But now I no longer had a grip. And I flinched when someone tried to shake hands as we ordinarily would, right hand extended. My right hand was curled into a claw, and refused to move. At Percy Jones, we were always on the lookout for a man's disabilities. Some, of course, were obvious. A guy with no legs, no arms, no hands. You didn't have to question how to approach a man like that. Other soldiers' wounds were internal, some were simply less visible, but a guy learned to be careful when greeting someone. A slight nod of the head was often sufficient.

But in Russell, as in most of America, people were accustomed to shaking hands. That day in Dawson's, I realized, maybe for the first time, the odd, awkward feeling people had when they stuck out a hand to greet me. And as best I could, I began extending my mostly useless left hand in return.

"Well, er . . . nice seein' ya," one fellow said as he looked as though he were about to slap me on the back like he would have done to the old Bob. His hand seemed to hit friction in midair, and he brought it down quickly to his side.

"Yeah, good to see you, too," I said.

It really was.

CHAPTER 21

The Good Doctor

I felt triumphant as my brother, Kenny, and I pulled out of Russell and headed for Percy Jones Army Hospital. It had been a good stay at home, and I was invigorated, ready and eager for the next round of treatments back in Michigan.

There had been some setbacks to be sure, and a few discouraging words. Like the guy downtown who said, "Poor Bob. He probably wishes the Germans had finished him off."

The guy probably meant well. It's just that some people have a hard time understanding how anyone could consider himself a whole person if he or she has a serious disability. I was having a tough time understanding that myself—but I was learning . . . the hard way.

During the remainder of 1946 and the first half of 1947, I kept looking for my miracle. I continued going to therapy, and tried every treatment the doctors suggested. Now that I was getting around a little better, I plunged into academic pursuits to pass the time. I met a friend, a fellow patient, George Radulescu, a former All–Big Ten football player who had suffered a broken neck in the war. George and I shared an interest in history. Together we studied the history of warfare, looking particularly at leaders and generals such as George Washington, Robert E. Lee, and Ulysses S. Grant.

I especially admired Dwight D. Eisenhower, a leader whose Kansas roots showed in his courage and character throughout World War II, and later in his conservative but progressive policies as president of the United States. Maybe I identified with his small-town upbringing, or his straightforward speaking of the truth, or his ability to forge alliances with people of differing opinions or positions; it's hard to pinpoint exactly why I liked him so much, but in a true sense, Dwight Eisenhower was my personal hero.

George and I would read, study, and debate almost every night until the nurse, Captain Louise Dobbins, came around and announced, "School's out, Lieutenant." We didn't mess with

Nurse Dobbins. She was a strict one, a tough cookie, but a great nurse.

My studies with George whet my appetite for further education. I began to think more seriously about going back to college when I was discharged.

Meanwhile, I was also learning to drive my specially designed General Motors automobile. The car developed a problem, however, and GM had to put a new motor in it, but I was getting around fairly well, driving with my left hand only. Because I was able to get out of the hospital for more extended leaves, I started scouting out the possibilities of visiting other doctors, looking for a second, third, or fourth opinion regarding what could be done about my right arm and shoulder.

The army and I had come to loggerheads; it seemed that we had radically different definitions of recovery. My definition was simple: I wanted to be put back together the way my body functioned prior to April 14, 1945. Of course, I recognized that I'd never be the same. I had lost a kidney, after all. And part of my shoulder was gone forever. But my arms and my hands were still intact; I wanted them to work. Surely, some surgeon somewhere, applying the miracles of modern science and medicine, could help me somehow.

The army had decided that physical therapy was the answer, and I was indeed making incremental progress. But at the rate I was improving, I'd be an old man before I could ever lift my hand, much less shoot a basketball or embrace a woman on the dance floor. Some days I spent hours trying to close two fingers on my right hand. It was extremely frustrating.

As the snow began to thaw in Michigan, I felt more confident about traveling greater distances. I consulted doctors in Wichita, Kansas City, and Denver. One week, I took a trip to Chicago to be examined by another doctor, to see if he thought he could do anything to help me. On April 13, the eve of the second anniversary of my being wounded, I wrote a letter to my parents, with the help of a Red Cross worker, to tell them what I'd discovered.

Dear Mom and Dad,
   I went to Chicago Monday to see Dr. Chandler. He recommended that a de-rotation osteotomy of the right humerus of approximately forty-five degrees be done. He explained it to

me this way: they cut the big bone in my upper arm and twist it outward forty-five degrees. He thought that by doing that I could have much better use of the hand because it would be away from my body, and that I could get it to my face, etc., much easier.

He thought that my hand would improve in time, but didn't think that my shoulder should be touched. I wasn't completely satisfied when I left his office, because he had spent such a short time with me. He looked at a couple of X-rays, and spent a little time examining my shoulder so I wasn't encouraged a lot after seeing him.

I'm going to see another Dr. in Chicago within the next two weeks. There's a Colonel here who used to be on Ward 23 with me. He is getting the appointment for me, and said that he might go with me, if he possibly could. He is a medical officer, lives in Aurora, Illinois. . . .

My car is running much better with a new motor. I have two hundred

miles on this one. If the phone strike doesn't end, someone may have to break down and write a letter. Not much else, so I'll say good-bye for now.

> **Bob**

My next appointment was with a doctor I'd first heard about from my uncle Lamont Jahn, who had served in the Medical Corps during the war. There, he met an unusually talented surgeon he simply called "Doctor K."

Dr. K was actually Dr. Hampar Kelikian, an Armenian refugee who had escaped to America as a boy with only a few dollars in his pocket and a small rug under his arm, when his homeland was ravaged by war. Dr. K's three sisters were not so lucky; they were killed in one of the senseless massacres that were all too common in their country early in the twentieth century. A brother of Dr. K's was killed during the Italian campaign of World War II. Dr. K understood the horrors of war all too well.

Arriving in Chicago in 1920, Hampar got a job on a farm outside the city limits. An extremely bright young man, he was also a dedicated, hardworking employee. So much so that his employer helped pay his way through college and medical school. When war broke out in Europe in 1939, Dr. K joined the Medical Corps. He quickly earned a reputation as a pioneer in the surgical restoration of arms and legs. That's where my uncle first heard about the talented surgeon who had a passion for helping American GIs.

By the time I learned of Dr. K in 1947, he had established himself as one of the top orthopedic surgeons in the country. Traveling to Chicago to meet with him, I allowed myself to believe that perhaps I'd finally found my miracle worker.

I sat waiting patiently for Dr. Kelikian in his large office. "Lieutenant Dole," the receptionist said, "the doctor will see you now. Please follow me." I dutifully followed the woman through the waiting room door, and started down an interior hallway. We hadn't gotten very far when a small dynamo of a man came rushing out to greet me.

"Hello, Captain Dole. Welcome. I am Dr. Kelikian. I am so glad you are here," the energetic doctor said in broken English. He was a bit hard to understand, and so excited to see me that I facetiously thought, **Maybe I'm his first patient.**

"How do you do, sir . . . and, ah, it's lieutenant. Lieutenant Dole."

"Ahh, yes, Captain Dole. Come, come," he said, nodding to me, while at the same time dismissing the receptionist. "Follow me."

A sophisticated-looking woman was seated in another waiting room.

"Please, you can wait," Dr. K said. "I need to see the captain. He came a long way."

The doctor was already moving toward an examining room down the hallway. I followed him into the room and sat down where he indicated. I tried to size him up as he pulled out my chart and scanned it as if reminding himself of an oft-read story. A short man, with prematurely graying curly hair, Kelikian was a bundle of energy and optimism. His eyes seemed to sparkle when he talked; he virtually exuded confidence. Extremely friendly, he was nonetheless brisk and businesslike. I could tell that Dr. K was not about to waste my time or his.

He examined me thoroughly, asking numerous questions as he felt around the bones in my arms, my hands, and my fingers. I struggled to understand him at first, but after a while I caught on to his broken English. He stared at my X-rays, then went back to my right arm, picking it up gently. When he let the arm go, it dropped lifelessly to my side. Dr. K raised his eyebrows and looked at me, but he didn't say anything.

After a while, he nodded toward a chair. "Sit. Please."

I had come to Dr. Kelikian looking for a miracle, and I couldn't wait to hear how the good doctor was going to make it happen.

He spoke kindly, with an accent, but without a trace of hesitation or reservation. I don't recall his exact words, but the essence of his message was unmistakable: "There will be no miracles. We can do surgery, probably several operations will be required. And I'm confident that we can restore thirty, maybe forty percent usage of your right arm. We can do some work on the upper arm and shoulder, but it is unlikely to ever rotate. You probably will not be able to lift your arm above your head, but you will be able to use it. We'll do all we can to give you back as much use of the arm and hand as is surgically possible. After that, it will be up to you to make the most of what you have."

In a nice way, Dr. Kelikian was telling me, "You won't be playing basketball. We'll do the best we can medically, but you have to stop chasing the rainbow. It's time to accept your situation, it's time to grow up and move on with your life."

Dr. K's words stung. They were not the words I had traveled to Chicago to hear; nor were they the words I'd been expecting—although deep inside I probably wasn't surprised. Over the past two years, I'd grown accustomed to doctors telling me that it was just going to take a little more time. "Keep working at the physical therapy. You're coming along fine." I had grown familiar with the language and terminology associated with spinal and nerve damage and impaired functions of arms and hands. But Dr. K was the first doctor to really shoot straight with me. There was something about his transparent honesty, his demeanor, his frank assessment of the situation that caused me to accept his diagnosis—not just of my arms, shoulder, and hands, but of my life.

"Well, what do we do?" I asked. "Where do we start?"

Dr. K leaned forward in his chair, his eyes glued to mine. "We start by not thinking so much anymore about what you have lost. You must think about what you have left . . . and what you can do with it."

I nodded, not sure how I was going to do that, but willing to follow the doctor's orders.

Dr. K had another surprise for me: He refused to take any money from me for his services. Not a dime. He explained that when it came to helping American GIs, he ran his business according to need. If you could afford to pay, he'd charge you accordingly. If you had no financial resources from which to draw, he would waive his surgical fees. Although no longer an army doctor, he would operate out of a love for his adopted country, and in memory of his fallen family members.

I left Dr. Kelikian's office that day thinking, **Well, I'm not going to be a professional athlete, and I'm not ever going to be a doctor . . . maybe there's something else I can do . . .**

Dr. K scheduled the first operation in Chicago for June 3, 1947. Although he was offering his services free of charge, the fees at the Wesley Memorial Hospital, where Kelikian would perform the surgery, would have to be paid somehow. That created a problem. Although by now I had been promoted to captain, as Dr. K had insisted on calling me from the first day we met, my postwar salary was rather meager. And there wasn't much time.

When Chet and Bub Dawson heard about Dr. K and his offer, they were excited about the possibilities. At the same time, they were concerned that I might miss the opportunity due to lack of money. Chet came up with an idea. He placed an empty cigar box on the counter at the drugstore, with a small: sign, THE BOB DOLE FUND, and began soliciting donations. He and Bub put the first few bills in the box.

Chet was commander of Russell chapter 6240 of the Veterans of Foreign Wars, so he made a heartfelt pitch to the guys there. Phil Ruppenthal, another of my high school buddies, joined in the effort. Everyone in the chapter knew someone who hadn't come home from the war. I had, but not really. Not yet. More than anyone else, the VFW knew what a struggle I was having, and the men responded.

Soon others in Russell took up the cause. Over at the Home State Bank, Bub Shaffer took up a collection. Another sprang up across the street at Russell State Bank. The Banker family, whose son Dean had endured his own personal hell in a German prison camp, pitched in at their store, Banker's Mercantile. So did many of the other shops and stores in town. Some people gave a dime; others a couple of quarters. There were a few big donors; mostly, however, the cigar box was filled with a lot of small bills and change. But the people of Russell rallied around me, and when I needed their help the most, they dug deep and helped. All totaled, the people of Russell donated eighteen hundred dollars to "The Bob Dole Fund." That may not seem like a lot of money nowadays, but believe me, in 1947, it was a fortune.

It was enough to pay the hospital bill. Excited and profoundly moved, I called Dr. Kelikian and told him the good news. I'd be coming to Chicago for my operation.

Hope sprang up in my heart. **Maybe miracles do happen,** I told myself. Only later did I fully realize what a miracle had already taken place in the storefronts and on the street corners of my hometown.

CHAPTER 22

A Good Right Arm

Nervous, excited, and a little fearful—those were the prevailing emotions I felt as I headed toward Chicago early in June 1947. For so long, I had been hoping for some change for the better in my condition, and anticipating the operation that Dr. Kelikian had described, yet I was still a bundle of nerves.

On one hand, I was excited about the possibilities. This was the first real attempt to repair my motor abilities since that night in the Evacuation Hospital in Italy, back in April 1945.

At the same time, I was fearful—afraid to think what I would do if the operation was not a success. Where would I go, to whom would I turn, what would I do then?

At the recommendation of Colonel Alling, an army doctor at Percy Jones, I was given "sick leave" so Dr. Kelikian could perform the surgery in a "civilian" hospital. The technical term the army applied to Dr. K's efforts was arthroplasty. In fact, Dr. K's goal was much more ambitious than mere reconstructive surgery on my shoulder and arm. He first cut away the bone in my shattered right shoulder, literally removing the head of the humerus. He drilled a hole in the humerus and removed the damaged "ball and socket." Then he took a strip of fascia—muscle covering from my left thigh—and used it to reinforce my damaged arm. Then he put me all back together again.

In the recovery room, both Dr. Kelikian and I were optimistic. If this procedure worked, I might have some use of my right arm again. I returned to Percy Jones to recuperate, hoping against hope that when the bandages came off, my arm would be nearly as good as new.

It didn't happen. Even while resting at Percy Jones, it soon became obvious that something was wrong; my arm wouldn't move properly. We gave it another month or so to heal. After two months, there was little change.

Undaunted, Dr. Kelikian took me back into surgery on August 11, 1947. Following the surgery, I still couldn't do much, but at least my arm was hanging somewhat normally, though my right arm was now two and a half inches shorter than my left.

Although I could bend my arm a little higher than my waist, I'd never be able to do much with it. My hand was practically useless. I'd never shoot a basketball with that arm, or flick my wrist like a fly fisherman casting his line across the water; nor could I take a deep, sweeping swing like a golfer, or put my strong right arm around my future wife and pull her close to me as we drifted off to sleep. Not with that arm. I'd never be able to toss my future daughter into the air above my head and see that look of sheer exhilaration and trust in her eyes as she waited for me to catch her. No, the arm had bones, muscle, and sinew in it, with blood pulsing through its veins, but for the most part it was a dead part of my body.

Increasingly, the fingers on my hand wanted to close into a ball. I struggled constantly just to keep my right hand open at all. In a third operation, on November 8, 1947, Dr. Kelikian attempted to implant some muscle and tendons in my right hand. The operation helped a little, but not much. My fingers continued to operate unnaturally, going in whatever direction gravity or motion took them.

I discovered, though, that I could now use them for something. I could roll up a piece of paper and wrap my fingers around it, giving the impression that my right hand was doing something. People who would ordinarily rush up to shake hands with me would now slow down when they saw that I had something in my hand. **Gee, that must be an important document, because he's clutching it so tightly.** Besides giving the impression that my right hand was "occupied," wrapping my fingers around the rolled paper also helped give my hand a normal-looking shape. I eventually discovered that by holding a pen in the hand, grasping it between my thumb and forefinger and wrapping the remaining three fingers around the shaft of the pen, I made the hand appear normal. For nearly six decades now, if you've seen me on the political campaign trail, in the U.S. Senate, in an interview, joking with Jay Leno on television, in a commercial, visiting with our troops, or anywhere, you've perhaps noticed that the pen is always there. It has become a part of my right arm.

In the years to follow, Dr. Kelikian would operate on me four more times—always for free, never charging me a penny—with limited success. Years later, when somebody asked Hampar Kelikian why he didn't give up working on me, he would respond in his still broken English, "This young man . . . had the faith to endure." To me, Dr. Kelikian's comment was one of the greatest honors I've received in my life.

Some people might regard Dr. K's efforts as failures. Not me. Perhaps from a strictly medical standpoint the operations were less than successful. There was just too much damage done to my body, particularly to my spinal cord, that day on Hill 913. Maybe the same type of wounds today would be easier to treat. I'm constantly amazed at the miracles I see at Walter Reed Army Medical Center in Washington, D.C. Maybe if my injuries had occurred today, and if Dr. K were living—he died in 1983—he could have performed a medical miracle.

Instead, he performed a different kind of miracle. He inspired within me a new attitude, a new way of looking at my life, urging me to focus on what I had left and what I could do with it, rather than complaining about what had been lost and could never be repaired. He encouraged me to see possibilities where others saw only problems. It's an attitude that has served me well over the years, and I will always be grateful to Dr. K.

He instilled confidence in me despite my limitations. Maybe that's why I accepted when my sister Norma Jean and her boyfriend, Tom Steele, a navy veteran, asked me to be a part of their wedding that fall. Beyond that—they asked me to be the best man.

Trinity United Methodist Church was packed with people that day. Weddings in Russell were still considered big deals, family affairs, and our extended family included the whole town. I was a little concerned how we were going to handle the ring ceremony during the wedding. As the best man, it was my responsibility to hand Norma Jean's wedding ring to her groom when the minister requested it. With my lack of feeling in my fingers, that could take a while—especially if I had to reach into my pocket and pull it out. Without looking, I couldn't tell the difference between a quarter and a ring, much less pull one out of my pocket.

The new minister at Trinity, Dr. Carl Eklund—Pastor Jenkins had been transferred to another location—helped me by slipping the ring onto one of my fingers before the wedding, so I'd have it ready. When the time came, the preacher simply reached over and slid the ring off my finger, and we never missed a beat; the congregation never noticed. All they saw was a big brother's pride in his little sister. And Russell, Kansas, knew that I'd finally come home.

After the third operation, I returned to Percy Jones on November 15, and continued a program of extensive physical therapy, trying to regain strength and motion in my arms and fingers. I spent the next several months learning how a naturally right-handed person could function as a "leftie." Percy Jones had programs for everything, including lessons to teach me how to write with my left hand. Every so often, I'd find myself getting discouraged, but whenever I got down, I'd recall Dr. K's words. Every day I was reminded that there were plenty of injured soldiers at Percy Jones Hospital and others around the country who were a lot worse off than I was. One such reminder left an indelible impression on me.

I could walk much better now, so I often went outside to smoke a cigarette. One blustery winter day, while I was waiting to be transferred to another part of the hospital for therapy, a buddy and I were outside when I noticed that a car had slid off the road, into a snowbank.

The driver, a lieutenant, climbed out of the car with the aid of a cane.

He called to us, "Hey, can you guys give us a hand?"

Now, that was a loaded question, but rather than try to explain, my buddy and I waded into the snow to offer whatever assistance we could. While the lieutenant gunned the motor, my buddy and I tried to push the car out of the snow, to no avail. My buddy was freezing, so he traded places with the lieutenant for a while. After two more unsuccessful attempts to extricate the car from the drift, the lieutenant suggested that I get into the car and warm up for a few minutes. While I was warming myself, he got out and began digging all around the rear wheels of the car with his cane.

Plenty warm, I got out of the car, ready to push again. I looked down at the lieutenant's feet, buried in the snow. "Aren't your feet cold?" I asked.

He looked at me and smiled. As he did, he thumped first one leg then the other with his cane. He didn't need to say a word. The sounds of wood against wood or some other hard substance indicated that both of his legs were artificial.

We eventually got the lieutenant's car out of the snow, but his face and that sound stayed with me long after he had driven away. The incident was a graphic reminder to me that when you start thinking about how tough you have it, there's always somebody else who could be dealing with a lot more.

On March 1, 1948, I was transferred to Ward 19 of Percy Jones Hospital, where the doctors set about trying to help me overcome the tremors and spasms that racked my body. I'd be walking along normally, when suddenly my body would begin to quiver. At other times, I'd try to raise a glass to my mouth, and spasms would strike, causing me to spill the contents all over myself. Sometimes my legs would give out and I'd have to sit down quickly. At other times I'd just stand there and shake, unable to move. We tried everything the docs could think of, but nothing seemed to help.

As a last-ditch effort, the doctors began treating me with curare in oil, some sort of South American snake venom with an oil chaser.

Really. It was a precursor, I suppose, to modern-day botox. Every four days, I ingested one dose of snake venom, in hopes that it would calm the spasms and tremors. It didn't. I took a short break from the regimen, and then tried another round of snake venom, again taking one dose every four days. There was no significant improvement in my condition, although the curare gave me a good excuse anytime I wanted to be surly.

That spring, I was in the mess hall one day, with my arm in a sling following an operation by Dr. K, when I noticed an attractive brunette at a table nearby having coffee with some other young women. I recognized the woman as an occupational therapist, a vital part of Percy Jones Hospital's efforts to help soldiers recuperate.

Occupational therapists were not career counselors, though I suppose that was part of the long-term plan. The army discovered early in the war that wounded soldiers responded better to all their treatments, physical as well as psychological, if they felt that they were accomplishing something, rather than simply lying in bed biding their time. A pioneer program was implemented that gave recuperating soldiers the opportunity to express themselves through arts, crafts, music, and other forms. Certainly, for some soldiers, the program helped them develop new skills that could eventually be used to make a living. For most of us, however, occupational therapy was mostly about reigniting hope.

For instance, as an integral part of therapy, I'd go along with other wounded soldiers to crafts class, where we'd work for hours making a design on a leather belt, a wallet, or some other silver, leather, or ceramic project. The occupational therapists constantly encouraged the soldiers: "Oh, that's beautiful. You've done a marvelous job on that belt." Looking at it, I knew it was just an ugly piece of leather with a bunch of markings on it, but the occupational therapist wouldn't hear of it: "Oh, no, that's wonderful! Let's send it home to your mom."

The therapists were right. What mattered, of course, was not how pretty a craft or project looked, or even how functional it was; the goal was to accomplish something. Yes, you may have some disabilities to overcome, but you are not a wasted human being; you are not going to sit in a wheelchair on a street corner in Russell for the rest of your life. You can do something. You can make something of yourself. (I can never adequately thank the occupational therapists at Percy Jones, especially for putting up with me.)

Nowadays, occupational therapists engage in a multitude of programs, helping disabled individuals develop new career skills, attitudes, and habits, as well as working with learning disabilities and other long-term care situations. But in 1948, the fledgling occupational therapy program in our army hospitals was pretty much dictated by the needs of wounded soldiers to do something productive with what remained of their lives.

Phyllis Holden understood that. Maybe that's why the pretty young graduate of the University of New Hampshire had signed on at Percy Jones Army Hospital. She had a heartfelt desire to help not just in the war effort, but also in the effort to help broken human beings rebuild their lives. And she was good at her job, too. Before joining the army, she had served in psychiatric wards, working with drug addicts, alcoholics, and schizophrenics. There wasn't a hint of fear in her. She went about her job teaching new rehabilitative skills with a missionary's passion.

As Phyllis recalled that day, when she saw me walk by in the mess hall with my arm in the sling, one of her friends commented, "Oh, there goes that poor Bob Dole. He doesn't have long to live, you know."

Phyllis's heart was touched. "Oh, that's so sad."

Actually, I was doing fairly well. By spring 1947, the doctors at Percy Jones and I were coming to the same conclusion: There was little more they could do for me. It was written in my official record, "It is felt that no further treatment is indicated for this patient, and that as far as regaining strength and function, he can carry out this type of exercise on his own."

I couldn't have said it better myself.

But being honorably discharged from the army for medical reasons is not as simple a matter as one might think. I was examined and re-examined by various doctors, and then both the doctors and I had to go before the official Army Retirement Board at Percy Jones Army Hospital to prove that I qualified for early retirement. It took a lot longer to get out of the army than it did to get in.

I didn't mind the wait. In fact, it proved to be quite interesting. One night, shortly after I'd seen the attractive brunette in the mess hall, I went to a dance at the officers club—one of the many activities the hospital provided for its patients—and there she was.

Phyllis and her friends had actually decorated the hall for the dance in a Heaven and Hell theme. Hell was pictured with flames and devils, and located in the bar area. Heaven included painted clouds and angels, and was located on the dance floor. A guy didn't have to be a genius to figure out that inference. I worked up my courage, walked over to where Phyllis was sitting with a group of friends, and asked her to dance.

"Why, sure. I'd love to," she said with a warm, engaging smile.

We walked out onto the dance floor and I took her right hand in my left hand, then awkwardly attempted to put my right arm around her shoulders as the 1940s style of dancing dictated. That wasn't going to work. I had to settle for putting my right arm around her waist. Phyllis understood; and she didn't mind. The good part about not being able to extend my arm to put it around her shoulders was that we had to dance closely. I didn't mind that.

Phyllis and I connected immediately. She had dark, sparkling eyes and a quick smile. She loved to laugh and had a great sense of humor—she even caught most of mine, which made me like her all the more.

At some point our conversation turned naturally to how I had gotten to Percy Jones in the first place. I gave her the broad-brush story about my experiences in Italy, and she seemed sad. Then I told her about Dr. Kelikian and his prognosis for my future. Phyllis was delighted to learn that her friend had been wrong; I wasn't planning on dying anytime soon.

That night, after the dance, I couldn't get Phyllis out of my mind. I woke up thinking about her the next morning. She was just such a compassionate and caring person. Two days after the dance, I called her at nine-thirty at night. "I, eh, was . . . well, I'm going over to Hart's Hotel for a cup of coffee and . . . well, I was wondering if you'd like to go along."

Phyllis caught on to the fact that I was asking her out on a date, kinda. "Sure, when would you like to go?"

"How 'bout now?"

"Ahhh, okay. Let me get my coat, and I'll meet you at the front door."

Phyllis and I went for coffee and talked well into the night. It was the first of many such meetings, when we talked and talked. She told me all about her upbringing, and I told her about my family, about growing up in Russell, Kansas, about Dawson's Drugstore, and the cigar box in which my friends had collected money so Dr. K could operate on me. Russell sounded like a foreign country to a New Hampshire girl, but Phyllis seemed intrigued nonetheless. We talked about my plans to be a doctor, and my hopes to be a great athlete one day, dreams that went down the drain along with the water in the Percy Jones therapy whirlpool. I told her about my late-night studies with George, which had rekindled a desire in me to go back to college, maybe to study law this time. Lawyers didn't have to use their hands or arms, just their heads. I thought I could do that.

Phyllis was extremely supportive. She never spoke condescendingly, as if my goals were unattainable because of my disability. Quite the opposite, she treated me as if I were any other normal guy, with two good arms and a fully functioning body, despite the spasms and tremors that rocked me occasionally, even when we were together. Best of all, Phyllis seemed to like me for who I was as a person, not what I could or could not do.

A few weeks later, I had another appointment with Dr. Kelikian in Chicago. I asked Phyllis to go along. The doctor must have noticed the way Phyllis and I looked at each other. After my exam, as we were preparing to leave, I told Dr. K, "I've been thinking of going back to school, to study law. What do you think, Dr. K?"

The surgeon shrugged his shoulders. "Why not?" Then Dr. K looked toward Phyllis and his bright eyes twinkled a bit, as he said, "Take her along. She can take notes."

When I went home for a short leave, just after Easter, I couldn't wait to tell everyone about Phyllis. I managed to squeeze her name into nearly every conversation. With my parents, with Kenny and Dottie, and their new baby—"You know, I think Phyllis would probably be a good mom, too, just like you, Dottie." With the guy over at the drugstore, anywhere that I went, I babbled like a fool in love.

Could it be?
Me?
Looking back, I wonder now how much of me feared that if I didn't hurry and ask Phyllis to marry me, I might lose her. That had happened to me before, and now here was this lovely, fun, energetic, compassionate woman,

who seemed to like me a lot and was totally comfortable dealing with someone with a disability . . . maybe we were a good match.

When I returned to Michigan a few days later, I couldn't wait to see her. But when we got together, Phyllis seemed upset. Her normally vibrant face was sullen, and her eyes were downcast. She had received a report back from the army listing several allergies that had shown up on her recent tests.

"Well, the army'd probably let you out," I said, "if you were to get married . . ." Phyllis looked up abruptly, but didn't say anything. After a few moments of silence, I continued, "Do ya think you could live in Russell, Kansas?" Like my father before me, I wasn't too good at love out loud. But Phyllis understood the question.

Her answer was "Yes!"

Now that I'd asked Phyllis to marry me, and she had accepted, we had to deal with the matter of telling her parents. I knew my parents and family members would be thrilled; they saw what a difference Phyllis had made in my attitude, and they were predisposed toward liking her before they'd ever met her. But Phyllis's folks—now, that was a different story.

When Phyllis informed her mom that she and

I were getting married, Mrs. Holden hit the roof. "Phyllis, no! You can't be serious," she said. Phyllis had told her about my being wounded in Italy and paralyzed for nearly a year, but that I'd made a remarkable—almost miraculous—recovery. "Phyllis, stop to think about it. You yourself said that he can't even button a button. How's he ever going to earn a living? How will he be able to support you?"

"He's going back to college, Mother," Phyllis said. "Send out the invitations."

I had my last medical examinations at Percy Jones Hospital on May 12 and May 14, 1948. Being examined one more time by the various departments was similar to going around to all your teachers in school, asking if they would sign off on your being absent from school the next week. Only this time I wasn't coming back. The doctors' official report about me stated:

> On 14 May, 1948, this patient was seen by Major Edward A. Ricketts of the Medical Department who gave the patient clearance for the previous pulmonary infarct. This patient received orthopedic clearance

by Colonel Rylander, Chief of Orthopedic Service on 12 May and his notation was that the patient was incapacitated for any further duty. On 12 May 1948, the patient was seen by Captain Joseph D. Picard of the GU Service who gave the patient GU clearance and a notation that no further difficulties were expected. . . .

The report concluded by recommending that I appear before the Army Retirement Board for further action. On May 26, 1948, I appeared before the board, and they officially pronounced that I was pretty much useless to the U.S. Army from then on. It was noted in my military records that during my time in the army, I had earned two Bronze Stars with an oak-leaf cluster and two Purple Hearts. My official retirement would take place on July 29, 1948, when I would be retired at the rank of captain in the United States Army.

I had spent the past thirty-nine months in army hospitals. I was ready to reclaim the years I'd lost and get on with my life.

CHAPTER 23

New Ambitions

By the summer of 1948, many men of the 10th Mountain Division who had fought on or around Hill 913 were already making their marks in civilian society. Ollie Manninen, the tough, athletic runner who earned a Silver Star for valor in battle that day, the man who first dragged me out of the sights of the German machine gunners, was training to compete in the 1948 Summer Olympics as a cross-country runner. Devereaux Jennings would go on to the Winter Games at Saint Moritz, competing in the skiing events as a downhill racer.

A number of the guys who had trained at Camp Hale, and fought in places such as Mount Belvedere and Castel d'Aiano in Italy, went back to Colorado to have an impact on the ski community there. Sleepy, snowbound towns such as

Aspen and Vail owe a great deal of their growth to the skills and energy of members of the 10th Mountain Division who went back home to start a whole new phenomenon, the luxury ski resorts in that part of the country.

And one fellow, Bill Bowerman, took a bit of the expertise he'd learned by trekking up the mountains of Italy, and he and a partner, Phil Knight, started a shoe company, offering products with good traction and comfort. The Nike shoe company has done rather well.

While many of the other guys were already changing their world, I was just now entering mine. On Memorial Day weekend, Phyllis and I drove from Battle Creek, Michigan, to Concord, New Hampshire. My mom and dad arrived a week or so later, and on June 12, 1948, Phyllis and I were married in Saint Paul's Cathedral. After the wedding, I drove my bride to Lake Winnepesaukee, in New Hampshire, where we enjoyed our honeymoon in a rustic cabin along the lake. Following the honeymoon, we headed west on Highway 40 toward Russell, Kansas, to spend some time with my family.

We weren't planning to stay too long, though. The doctors at Percy Jones told me that because of the large amount of blood thinners I'd been subjected to, I would likely do better living in a warmer climate for a while. I had applied to and been accepted at the University of Arizona, where, thanks to the GI Bill—legislation that paid for honorably discharged veterans to go to college—I was able to go back to school debt free.

Phyllis and I decided to spend the remaining part of the summer in Russell until it was time to head for Arizona. It was an unusual feeling coming home this time, and bringing a wife along with me. But Mom and Phyllis hit it off wonderfully. Phyllis admired Mom's constant working to meet the needs of her family, and Mom admired Phyllis's willingness to watch and learn. After all, my mom was one of the best cooks I'd ever known, and a fastidious home-maker. Everything about our little home on Maple Street was always immaculately clean. Our clothes were washed and pressed, and even a tiny strand of hair on my shirt would drive me to distraction. "Oooh, get that off of me," I'd say. For Phyllis to step into my life and hope to do everything that my mom had done as well as Mom did was a tall order.

Phyllis knew how to cook, but she wanted to learn how to make some of my favorite dishes the way Mom made them. That seemed easy enough, but my mom cooked by taste rather than textbooks; she had few recipes and rarely measured anything, adding a pinch of this or a spot of that. Nevertheless, Phyllis had her notepad out, trying to scribble out recipes and write down the various ingredients that Mom used to make everything taste so good, although Mom never did the same thing twice. But she appreciated Phyllis's desire to please me.

The one big difference between Phyllis and my family was the way they treated me. My family coddled me, humoring my every whim since I'd been wounded. Like most families or close friends of disabled people, they didn't know what they should do, or not do; nor did they know how to relate to me. They didn't know how much assistance I really needed and what they should allow me to do for myself. Consequently, they stepped around me like a piece of delicate porcelain, as though I'd crack with the slightest bump, jostling, or displeasure.

Phyllis, however, knew that I needed to do things for myself, that, indeed, I **wanted** to do things for myself. One of the things that I appreciated so much about Phyllis was that she refused to baby me. Instead of waiting on me hand and foot as Mom had done, she'd often say something such as, "Get it yourself," or, "Now, Bob, you can do that. You don't need me to pick up your socks."

Phyllis wasn't being lazy, mean, or insensitive. Quite the contrary. Out of her concern for me, she refused to treat me as a cripple; she knew that the best way I could be happy was to do things for myself. As much as it hurt her to do so sometimes, she'd allow me to struggle on my own rather than jumping in and helping, just to make things easier for me, or for her.

My family members didn't always understand that. Once, for example, while we were home that summer, the family was gathered around the dining room table. I was holding a glass of iced tea in my left hand when the tremors struck. The glass slipped out of my hand and shattered on the floor.

"Oh, Bob, can't you hold on to anything?" Phyllis teased. A dead silence fell over the room, as my family members stared at her in disbelief. They couldn't imagine saying such a thing to me and were aggravated with Phyllis for doing so.

Mom swiftly grabbed some napkins and began sopping up the mess. "It's okay, Bob. It could have happened to anybody," she said, glaring at Phyllis.

But Phyllis understood me in a way that no one else did at that time. She loved me unconditionally, and she knew that if I was ever going to get better, I couldn't allow every little setback to be a stumbling block. To reinforce the idea that I wasn't a whole person, or that I couldn't make it on my own, would have been counterproductive. In my best moments, I didn't want people to make excuses for me or to feel sorry for me. I wanted to be a productive person. I didn't want to be a spectator; I wanted to be a player.

Eventually my family members came to appreciate Phyllis's positive influence on me, although it was rough going at first. But Phyllis helped me tap into the drive to overcome the obstacles in my life, something that was already there, as evidenced by my desire to excel in sports, to go to college, to become a doctor, and, now, to go back to school and, I hoped, get my law degree.

In September Phyllis and I packed up our Oldsmobile and moved to Tucson so I could begin taking classes at the University of Arizona in a liberal arts program. We rented a two-room apartment near campus, which, considering our meager belongings, was more room than we needed.

Phyllis was allowed to attend class with me, to take notes, and even to write out my answers on test papers from my dictation, but the work had to be all mine. Like many GIs going back to school after the war, I returned to the classroom with a fresh zeal and a renewed interest in my studies. Just getting by wasn't good enough anymore. Besides being older—I was now twenty-five years of age, and more mature than most of the younger students—more than ever I realized the value of a good education. Beyond that, I felt compelled to learn everything I could in the short period of time I would be there. Nobody had to urge me to study. Just the opposite. Phyllis often had to encourage me to take a break. "Bob, your grades are good. You're doing fine. Why do you want to study so hard? Why do you feel that you have to make an A?"

"I don't know how to study for a C or a B's worth," I told her. "I've got to learn as much as I can. I've got to get it all."

Maybe my tenacious study habits had something to do with the way I had to apply myself to overcoming my physical limitations, I don't know. To bring myself back to any physical normalcy, I had to give it everything I had; I guess that same attitude simply spilled over into my academic pursuits.

I studied constantly, often all night. Phyllis frequently went to bed without me, while I stayed up and tried to memorize the material I needed to know for class the next day. We probably both knew that such a pattern wasn't good for our fledgling marriage, but we felt it would be worth it, that it was a short-term sacrifice for long-term improvement. After all, we were together all day in classes, at the library, and at campus events. If we didn't get to spend much time in private conversation at home, well, that was just part of the price we were paying right now.

The weather in Tucson agreed with me, and when I did take a brief break from my studies, I was able to go out walking and even jogging a little in the hills surrounding the city. The physical strength in my legs was coming back more every day. Although I'd never regain the power, or even the shape of and tone in my legs that I once had, I could at least walk and run to some extent.

I was doing quite well in my classes at UA,

when one night at dinner, I felt an all-too-familiar pain in my lung. Before I knew it, I had nearly collapsed in my food. "Phyllis," I gasped. "Get me to a hospital, quick!"

I'm not sure which of us was more terrified. Phyllis hadn't really driven much since our arrival in Tucson, but somehow she found her way to an emergency room. The doctors wheeled me in on a gurney, and sure enough, I'd developed another blood clot on my lung. They kept me in the hospital for a week while they treated me with dicumarol to thin my blood. I knew the routine—nothing but bed rest for a while.

I had Phyllis bring my books and notes to the hospital so I could keep up with my studies. Fortunately, I was fine after a week, and the doctors allowed me to go home, which for us, meant back to school.

After completing one year at the University of Arizona, I committed myself to going into law school. But UA didn't have a law program at that time, so I immediately thought of returning to my old stomping grounds at the University of Kansas, in Lawrence. Since my latest bout with the blood clot, however, my blood needed to be tested and analyzed regularly to make sure it was clotting correctly, and KU didn't have a nearby lab in those days. The closest school to a sophisticated lab was Washburn University, in Topeka, which at that time was known as Washburn Municipal University. Phyllis and I decided to go there.

As I drove into Topeka, the Kansas state capital, I looked around the city in amazement. I had spent my first few months back in the United States after the war right there in Topeka, but I could barely recall ever being there.

"I was real sick the last time I was here," I told Phyllis as we drove into town.

We found a great little place to live, too—a three-room apartment with rich-looking dark wood trim, right across the street from the Kansas state capitol. The apartment complex was named "The Senate." Maybe that was prophetic.

I enrolled in a program that would lead to both an undergraduate degree in history and a graduate degree in law. Once again, as we had done in Arizona, Phyllis went with me to class to take notes, or else helped me to transcribe my left-handed scribbles into something legible each evening after classes. Fortunately, about that time, the Veterans Administration gave me a great tool to help in my education. It was a large, bulky, boxlike "Sound Scriber," a precursor to portable tape recorders. The Sound Scriber looked and worked a lot like an old-fashioned phonograph, and was about the size of one, too, roughly about the size of a modern-day computer printer. Similar to a "read-writable" CD today, the Sound Scriber recorded the sound directly onto five-inch spinning disks that could then be played back later, much like a record player, albeit with a much more scratchy sound quality. It was sometimes hard to hear my professors' voices for the scratches, but it worked.

Each day, I'd lug the thirty-pound machine to class, plug it in as close to the professor as possible, and insert the disk to record the lecturer's voice. Then at night, I'd replay those disks over and over, loudly, as I transcribed the lectures on the disks, scribbling with my left hand. I pity the neighbors in the apartment next door. I'd pace back and forth in our apartment with a book or notes in my hand, committing as much of the lecture material as possible to memory, including the various legal cases, citations, and rulings. I'd quote the material aloud until I could dictate it back.

Phyllis took a job in her occupational therapy

field at the Topeka State Hospital and at the State Rehabilitation Center for the Blind. Her work was a good outlet for her, as well as extra income for us. We didn't have much of a social life in Topeka. We were too busy working. Our only relief was an occasional drive over to Russell, nearly one hundred eighty miles away, to visit with my family.

Most of my professors allowed me to take my exams orally. For those who didn't, Phyllis accompanied me to the test. The two of us would sit off to ourselves, away from the other students, where I was permitted to dictate my answers to her, and she would write them on the test papers. We did the same thing later when it came time for me to take my bar exam, and Phyllis was as nervous as I'd ever seen her. She was worried that she wouldn't be able to spell the various confusing legal terms. Thanks to my aunt Mildred's twenty-five-cent incentives, I was still a good speller, and we made it through the bar exam with flying colors.

Often I'd stay long after class, discussing cases or principles with my professors. I wasn't interested in hanging out at the student union or going to campus parties. Like most of the men and women who had been in the war and were

now in school, I had a new seriousness about my
education. I wanted to learn more than mere
keys to success. I was searching for meaning. I
had a new appreciation for the American way of
law, justice, and government. I had seen firsthand in the bombed-out villages and the shredded, mangled bodies where less noble forms of
government could lead. Somehow, those memories removed the study of the rule of law from
the realm of the purely pedantic and philosoph-
ical, and put real faces on it.

There were off-campus lessons, too. Not far
from where Phyllis and I lived was a Phillips gas
station. An attendant named Shorty went out of
his way to assist me—doing everything from fix-
ing my car when it broke down, to miraculously
getting it started on cold winter mornings.
Shorty had probably never opened a law book in
his life, but his kindness and compassion helped
humanize "contract theory" and made even my
most boring subjects bearable because I constantly tried to apply them to real life—to everyday working people like Shorty and millions
more like him.

Of course, at Washburn, I was blessed with an
abundance of brilliant and wise professors—the
two traits don't always go together in the legal

profession, as they do in many others—men and women who challenged me to take what I'd learned from World War II and live up to the high principles and post-war resolutions I espoused.

One person who had a particularly profound influence on me was Beth Bowers, the law librarian at Washburn. Beth was only a couple of years older than I, but she was a strong, brave woman. Divorced, Beth Bowers was also a devoted single mother raising a child while completing law school in two years. Beth refused to accept complacency, let alone mediocrity, in the students with whom she rubbed shoulders. She constantly challenged my classmates and me to take an active role in politics. Her logic was simple. "You fought a war for freedom, to protect democracy in distant lands. How can you stand aside and allow it to decay on our own soil?"

It was Beth Bowers who first suggested that I run for office—the Kansas state legislature in 1950. It certainly wasn't the pay that piqued my interest. Our state representatives at that time received the whopping remuneration of five dollars per day, plus seven dollars per day in expense money during the biennial sessions. Nevertheless, I was intrigued by the possibility of learning more about the law firsthand. With the Washburn campus in such close proximity to the state capital, the Washburn law school already doubled as a think tank and employment agency for much of the Kansas state government. Many a Supreme Court law clerk bore the stamp of Washburn. My classmates often headed for the legislative galleries on the third floor of the capitol building for a taste of political science brought to life. I was right there and would be for at least another year; why not make the most of my time?

Besides, maybe Beth Bowers was right. I had experienced a lot of life in my twenty-seven years; I'd learned a lot and I'd overcome a lot. I came close to dying three times, facing death and living to tell about it. Why was I still alive? Maybe there was some bigger meaning to my life. Maybe there was something more that I was meant to do.

Calvin Coolidge told a story about himself as a child in Vermont. Coolidge said that when company came to visit, he'd often hide for as long as possible before going in to greet the guests. Standing behind the kitchen door, he could hear his parents entertaining, and he knew that eventually he would have to walk through that door and be sociable. As an adult, every time he went up to a voter to introduce himself, it was as though he were confronting that kitchen door again.

I can understand that feeling.

I felt like that during that first walk downtown past my neighbors with the words to the song "You'll Never Walk Alone" echoing through my mind. By then, I had begun to re-make myself physically, but more important, I had been re-formed emotionally and spiritually. I now knew that I had the inner strength to take on any challenge that might come my way, and those that I sought out.

Then, in 1950, when Beth Bowers encouraged me to run for political office for the first time, it was another mountain to climb, another door I felt compelled to go through. My entrance into politics certainly wasn't propelled by partisan fervor. The Russell County Attorney greatly influenced my decision to become a Republican. He told me, "Bob, if you really want to do something in politics in Kansas, you better declare yourself as a Republican."

"Really, why is that?"

"Because Republicans outnumber Democrats in Kansas."

I became a Republican, pragmatically at first, and then philosophically later on.

Over the years, I've climbed many hills and traversed many a valley. Going up Hill 913 and coming down on a stretcher changed my life forever. Then there were the figurative hills— something as simple as walking downtown, past the homes and eyes of my Russell neighbors, for the first time. That was a one-way station on the road to a hoped-for full recovery. Running for office would entail another kind of walk. Looking up the dirt road at a farmhouse on the outskirts of town, I could imagine someone on the other side of the door, the hasty introductions and the awkward request.

Still, there was nothing to be gained by delay. Everything that had happened since April 14, 1945, pointed me in this direction. **When you walk through a storm . . .** About to ask a neighbor for the first time to give me his vote, I realized I wasn't walking alone. Accompanying me was everyone who had made it possible for me to walk at all: Mom and Dad, my sisters and brother, Chet and Bub Dawson, Adolph Reisig, teachers and preachers and the banker who loaned me three hundred dollars and insisted I wear a hat, and KU friends and instructors, and army trainers, and buddies in and out of the foxhole, and merciful doctors and nurses and fellow patients who eased my suffering and infused me with the will to go on.

I walked with, and for, them. Approaching the farmhouse door, I saw the door to my future. Stepping onto the porch was to step into a new life, in Dr. K's words, making the most of what I had left. Alone? Hardly. Dr. K was at my side. And so were the guys who never made it back from the war, and who, half a century later, would have their sacrifice enshrined in bronze and stone in a city that would become my second home.

All this was unimaginable in the summer of 1950 when I knocked on that farmhouse door. It was opened by an elderly farmer who looked as though he'd just come in from a hard day's work in the fields. He pushed open the screen.

"Can I help you?"

He didn't know it, but he already had.

CHAPTER 24

Memorial Day

Nearly six decades had passed since Beth Bowers suggested that I go into politics. It was Saturday, May 29, 2004, Memorial Day weekend, and I found myself standing on a bright blue platform on Washington's famous mall for the dedication of the National World War II Memorial. It was a tribute long overdue to the 16 million men and women who had served in the armed forces during the Second World War. Millions more supported the war effort by changing their lifestyles at home, contributing in large ways and small, enduring everything from the rationing of gas and food products to the collecting of used tin cans to be made into munitions. Nearly half a million American soldiers had died, and another half a million were severely wounded on the battlefields of Europe and Asia, and in the Pacific. It was high time that we as a nation said thank you.

On that gorgeous spring day, our nation's capital was jammed with hundreds of thousands of visitors, many of whom had come specially to take part in the various Memorial Day festivities. At the National World War II Memorial site, I looked out on more than 150,000 people who had gathered for the official dedication ceremonies. Millions more were watching on television. Around me, several veterans of the war—including a number of Congressional Medal of Honor recipients—were sitting on the same platform in wheelchairs. Behind them was another row of honored vets dressed in suits and ties, but wearing various colored baseball caps bearing the logos of their branches of service in the military. To my left was Tom Brokaw, one of the early supporters of the memorial, and whose books and television broadcasts had pinned the label of "The Greatest Generation" on the men and women who fought the Second World War. Beyond Brokaw were other dignitaries, including former presidents Bill Clinton and George H. W. Bush; actor Tom Hanks, the national spokesman for the World War II Memorial Campaign; Frederick W. Smith, founder and CEO of

FedEx; Senator John Warner, who was always there when we needed help; General P. X. Kelley, chairman of the American Battle Monuments Commission; and President George W. Bush.

The suggestion for a World War II memorial came from Roger Durbin, a World War II veteran. Ohio Congresswoman Marcy Kaptur introduced the idea in 1993, and Congress authorized placement of the memorial within the Washington Mall area, with the site approved by the Commission of Fine Arts, the National Capital Planning Commission, and the Secretary of the Interior in 1995. Numerous potential sites were hotly debated, with an area known as the Rainbow Pool, at the east end of the Reflecting Pool between the Lincoln Memorial and the Washington Monument, finally selected. President Clinton dedicated the memorial site on Veterans Day, 1995.

On January 17, 1997, I was at the White House, where President Clinton honored me with the Presidential Medal of Freedom, the nation's highest civilian award. Established by President Harry S. Truman in 1945 to recognize individu-

als for their public service, the award was rein-
troduced by President John F. Kennedy, who
made it our nation's highest civilian honor.

To be included among the men and women
who have been awarded the Presidential Medal
of Freedom over the years was a humbling experience. Their company included presidents
Ronald Reagan, Jimmy Carter, Lyndon B. John-
son, and Gerald R. Ford, and other distin-
guished citizens such as Walt Disney, Clare
Booth Luce, Jeane Kirkpatrick, Arnold Palmer,
and Bob Hope. More recent honorees included
Colin Powell and Pope John Paul II.

After the ceremony, President Clinton
showed me a model of the proposed National
World War II Memorial. I was impressed. The
plan was for the government to "seed" the project, with about 95 percent of the costs to be
raised through private donations. Then President Clinton said something like, "And we think
we know just the man to do the job. Would you
be willing to be the National Chairman of the
World War II Memorial Campaign?"

"Well, let me think about it," I replied.

I didn't have to think long. How could I turn
down such a request from the president? How
could I refuse such a challenge; more important,

how could I miss the opportunity to honor the men and women with whom I had served during World War II?

Before announcing my decision, I went to see Fred Smith, who was in Washington on business. Fred was a graduate of Yale, an astute businessman, the CEO of FedEx, and a good personal friend. Five of his uncles had served in World War II, and Fred himself was a U.S. Marine Corps officer and a hero in the Vietnam War. I wanted to ask Fred to be my co-chairman, so I started into my pitch. "Fred, I gotta have help," I said. "I have this assignment to raise money to build the National World War II Memorial, but I don't know many of the new corporate leaders—"

Before I got any further into my spiel, Fred interrupted me. "Do you want me to do this?" he asked. "Does Bob Dole want me to do this?"

"I sure do," I replied.

"Okay, I'll do it." And he did. Fred worked for months, lining up calls to the top corporate officials in America. Many of them responded warmly. Not everyone, of course. "We don't give money to bricks and mortar," some of them said. "We give to causes that will help people."

Oh, okay. **Hmmm. Seems to me that the**

Allied victory in World War II helped millions of people . . . and many big corporations.

The job before us was made tougher because many of the business leaders who had lived through World War II had retired. Younger men and women occupied positions of authority, and some of them couldn't grasp the need for a World War II Memorial. On the other hand, CEOs such as Hank Greenberg, of the insurance giant AIG, were more than receptive to our requests for help. I must confess that one of the biggest disappointments was the lack of support from the Hollywood community, a group that had profited greatly from World War II movies. With the exception of Tom Hanks and Steven Spielberg, Hollywood refused our repeated requests for help.

The estimated price tag for the memorial was staggering—with construction costs of approximately $70 million, plus millions in management and design fees. Making matters worse, the costs went up almost every year. Nevertheless, our volunteers worked tirelessly for several years raising the needed funds, bringing in more than $197 million (including the government's seed money and interest) from more than

600,000 contributors—from businesses, individuals, and families who wanted to honor their loved ones who had served in the war. After everything was complete and paid for, we had a surplus of $15 million, which will be used to maintain the memorial.

While I had nothing to do with the actual design of the monument, I was pleased with it after certain modifications were made. The designers came up with a marvelous plan to maintain the esthetic integrity of the Mall, by placing large portions of the memorial below ground level. They incorporated the Rainbow Pool right into the design and created a magical public space.

Just about the time everything was coming together, we were hit with legal wrangling by groups who said that the Mall itself was sacred ground and shouldn't have any more monuments built on it. One group called itself Save the Mall.

"We already saved it once," I said. "We saved it and everything else in World War II."

The lawsuit was dismissed, but it took two or three years to fight through the appeals. Finally, after an eleven-year process, the monument commemorating the service and sacrifice of an entire generation was ready to be dedicated.

Two days before the memorial opened to the public, I escorted fourteen members of Congress who were World War II veterans (including my fellow Percy Jones alumnus Senator Daniel Inouye) through the memorial. Nearly all of them were profoundly affected. Most wiped tears from their eyes as we made our way between the great stone arches commemorating the Atlantic and Pacific Theaters.

Now, as I prepared to speak that Memorial Day weekend 2004, I saw a similar reaction among many older men and women in the audience. At least fifty-five Medal of Honor winners were on hand—good old guys who were once good young guys. Proud of their country, they let their tears flow freely on this day.

Tom Brokaw, Tom Hanks, and Fred Smith delivered poignant messages, reminding the audience of the selfless courage exhibited by the greatest generation, as ordinary people were called upon to do extraordinary things—and they did them, without fanfare or thought of notice or reward. It was obvious that the crowd was emotionally touched by the ceremony, as was I. Then it was my turn to speak.

"What we dedicate today is not a memorial to war," I observed. "Rather it is a tribute to the physical and moral courage that makes heroes out of farm and city boys, and inspires Americans in every generation to lay down their lives for people they will never meet, for ideals that make life itself worth living."

Indeed, looking back over eighty-one years (and still counting), those were the kinds of ideals that shaped my life. I'm often asked, "Where did you find the strength to overcome the obstacles you've encountered? What gave you the courage to get back up in life, after being blown off your feet on that rugged hill out there in Italy?"

My short answer is usually, "My parents." Mom and Dad instilled in me the values and principles that have guided my life all these years. Of course, the longer version of that answer must include my entire family, along with the good friends who were there for me when I needed them; and as I said when I announced my candidacy for president in 1996, at an early age, I put my trust in God rather than the government, and never confused the two. Maybe that's why I've always had a future orientation— I've always believed that there's more for me to do, places to go, people to see, experiences to encounter. I never wanted to slow down, not at seventy-three years of age, when I ran for president, and not now. For half a century or more, I've had a sense of making up for lost time. Even now, I find it difficult to sit still for long and "act my age." I've gotta go! In 1998, for example, I logged over 200,000 miles traveling to encourage humanitarian efforts. I hope to do more in the days ahead. When you stop dreaming about the future, and quit looking for new projects to do, you dry up like a prune, and life becomes boring. I want to keep going and growing.

Two qualities that I've learned through my war experience and the subsequent disabilities I've lived with for the past sixty years are patience and adaptability.

Patience is an acquired trait, and I've spent a lifetime impatiently trying to acquire it. Most of us want patience, and we want it now. But few things will cause you to stop and focus on the moment, as well as on the big picture, more than not being able to get out of bed for six months, not being able to feed yourself for more than a year. You come to measure progress in small, incremental steps. And you learn perhaps the toughest lesson in life: to wait. But not just waiting idly; no, to wait, as Dr. Kelikian said, with a faith to endure—believing that you have a future.

Perhaps that's one reason why I sometimes didn't try to rush legislation through the Senate. In fact, if I had any skill as a legislative leader, I probably possessed an instinctive knack for knowing when the time was right to make a move, or to attempt to make a good deal. Part of that patience, I'm sure, I learned at Percy Jones Army Hospital, while I was trying to get my fingers and toes to move.

The second habit I owe to my war experience and its aftermath is adaptability. The army is a good place to learn that trait under any circumstances, but when you are wounded, and suddenly your life's dreams get blown away, you must adapt.

I've always wanted to speak to the present as well as future generations. When I lost the presidential election in 1996, the American people gave me opportunities rarely extended to other candidates who have lost elections.

Just three nights after the end of a grueling

election campaign that had me crisscrossing the country for months, I accepted an invitation to appear with David Letterman on his television talk show. David was kind and gracious. With an impish look, he posed his first question to me: "Bob, what have you been doing lately?"

The crowd cracked up.

"Apparently not enough," I said. The audience howled again.

"You look great," Letterman said. "You look like you're not tired, and you've already bounced back and are ready to go."

"I'm ready," I deadpanned, "but no place to go."

It was the start of a whole new career for me. Playing up the image of the downtrodden, also-ran was great fun. For instance, when actress Susan Lucci did not win an Emmy Award for the umpteenth time she was nominated, Jay Leno's producers called and asked if I'd come on **The Tonight Show.**

"Susan Lucci lost for the seventeenth time," Jay told the audience in his monologue. "Seventeen times going for the big prize, and losing! She's like the Bob Dole of daytime television. It's unbelievable. It is so . . ."

Just then, I walked out from behind **The**

Tonight Show curtain, to the enthusiastic applause and cheers of the audience. Jay, of course, feigned surprise.

"How are you, Bob?" he asked as the crowd quieted a bit.

"I'm fine," I said, pretending to be irritated. "But Bob Dole doesn't like that joke." The crowd crowed wildly, instantly getting the joke as I poked fun at my oft-parodied practice of speaking of myself in the third person. (I don't know how or why I ever started doing such a silly thing in the first place. False humility, I guess.)

"Bob Dole is fed up," I continued. "Bob Dole has had five years of this, Leno . . ." and then I broke up laughing, too.

Maintaining a healthy sense of humor is a key to overcoming any setback in life, even when your setbacks are extremely public. In my speech at the White House after accepting the Presidential Medal of Freedom from President Clinton, just a few months after I'd lost the election to him, I began as though taking the oath of office.

"I, Robert J. Dole," I paused as the august crowd of political leaders and members of the press immediately caught on, and roared in laughter.

"Do solemnly swear," I continued without breaking a smile, to gales of laughter. I looked up as though surprised.

"Sorry, wrong speech." The crowd roared again. "But I had a dream . . ." The audience chuckled at my allusion to Martin Luther King, Jr., "that I would be here this historic week receiving something from the president . . . but I thought it would be the front door key to the White House." I looked over and President Clinton himself was redfaced with laughter.

Shortly after that, I was featured in an award-winning thirty-second Visa debit card commercial that ran during the Super Bowl. The premise for the commercial had me returning to Russell, Kansas. With pastoral music playing in the background, chirping birds, and scenes of me walking down Main Street, the commercial featured me speaking as though making a good-bye speech.

"And so, after thirty-five years in the service of my country, I return to Russell, Kansas, to the place I called home . . . to the small town that nurtured and defined me, to my friends and my family . . . to that very special place where they always called me by my first name. . . ."

Meanwhile, under my voice-over narration

were cut-in video clips of people saying how well and how long they'd known me.

The commercial then depicted me supposedly having lunch with all my friends at one of our local Russell restaurants.

"Great lunch," I said, as I approached the counter to pay. "Will you take a check?"

"Of course, Bob," the friendly woman behind the counter answered. "Can I see some ID?"

My face registered shock, my countenance dropped, and her tone turned cold. "Driver's license? Passport? Military ID? Voter's registration card . . ."

I looked at the camera and lamented, "I just can't win."

I later found out that more people saw me in that one television commercial during the Super Bowl than the number who watched the Clinton presidential inauguration.

Throughout much of my political career, many people didn't know that I had a sense of humor. Some people accused me of being a tough guy, the hatchet man for Gerald Ford as his running mate for the presidency in 1976; I suppose I was. Others have called me "Mr. Grinch," "Mr. Grumpy," "Mr. Gridlock," and other not very complimentary titles. To them, I was a partisan gunslinger in the Senate. But I've never seen myself in quite that way. I've tried to treat everyone with good humor and respect, whether they were a custodian or a committee chairman.

I have seen myself as tough. I was a leader in the Senate, and if you're going to be a leader you have to be tough at times, but not nasty. I was the guy carrying the flag, and I knew that if I didn't want to lead, I could pass the flag off to somebody else. But as long as I was the guy out front, my job was to pump up the troops.

On my office wall in Washington is a framed copy of the statement that General Dwight D. Eisenhower planned to release to the press and the world if D-day had been a failure. It's one of the best definitions of leadership I have ever seen. "Our landings have failed," wrote Ike, "and I have withdrawn the troops. My decision to attack at this time and place was based upon the best information available. The troops, the air, and the navy did all that bravery and devotion to duty could do. If any blame or fault attaches to the attempt, it is mine alone."

Think how many of our country's problems would vanish overnight if we could just get those words straight: The responsibility is mine and mine alone. In the final analysis, that's what great leaders do, not just in the Senate but also in daily life. They face life without flinching. They make the tough decisions. They live with the consequences whether good or bad. They make the most of what they have. That's the kind of leader I have always tried to be.

These days, I'm sometimes asked what accomplishments I am most proud of during my tenure in the Senate. My answer sometimes surprises people. I tell them that the things I reflect on most fondly are not party or personal victories, but things like saving Social Security and passing the Americans with Disabilities Act. These were accomplished by working together closely with both Republican and Democratic friends.

My first speech on the floor of the U.S. Senate, on April 14, 1969, was in support of a task force to improve the lives of disabled individuals:

My remarks today concern an exceptional group which I joined on another April 14, twenty-four years ago during World War II. It is a group no one joins by choice. As a minority, it has always known exclusion, maybe not exclusion from the front of the bus, but perhaps from even climbing aboard it.

From there, I went on to outline a broad range of programs to improve the lives of the nation's disabled. My own disability has sensitized me to the needs in the disabled community ever since.

Disappointments? Oh, yes. I've had a few. On a personal level, Phyllis and I divorced in 1973, the same year Richard Nixon embroiled the country in the Watergate scandal. I had been the chairman of the Republican National Committee at the time of the break-in that eventually brought down Nixon's presidency.

Explaining the explosive charges associated with Watergate to my constituents back in Russell was difficult enough. Explaining Phyllis's and my divorce was no easier, especially when many people still believed the fairytale about the wounded war hero being nursed back to health by his occupational therapist. That wasn't the case, and Phyllis was always quick to point out that I was well on my way to a healthy recovery by the time we met. But that was the myth. And myths are hard to overcome.

Nevertheless, Phyllis and I have remained good friends over the years, and I will always be grateful to her for being there for me at a time in my life when I felt so fragile. And for giving me one of the greatest joys of my life, our daughter, Robin, who I am happy to say has never held a government job.

I remarried in 1975, and have remained happily married to Elizabeth Hanford Dole for more than thirty years. Elizabeth is a remarkable woman who has brought much happiness to my life, and who, through her service in the cabinets of presidents Reagan and George H. W. Bush, and as president of the American Red Cross, has earned a reputation as one of the most admired women in the country. One of the proudest days of my life came on January 7, 2003, when I escorted Elizabeth to her swearing-in as the U.S. senator from North Carolina after the voters there elected her to succeed the venerable Jesse Helms. The ceremony was conducted by Vice President Dick Cheney in the historic chamber where I had spent more than twenty-seven years, including the longest run ever as the

Republican Leader in the Senate. It took a while to get used to addressing my wife as Senator Dole, but I'm happy to do it.

Ironically, six years after Bill Clinton and I faced off in a battle for the White House, we found ourselves reunited—as members of the "Senate Spouse Club." I often joke that we are going to run against each other for president of the club—and this time, I'm going to beat him.

Have there been some things that I wanted to do but couldn't because of my war wounds? Sure. I always enjoyed running, and I missed out on running a marathon, something I always wanted to do. While I could jog again after several years of rehabilitation, I tired easily, and my stamina wasn't up to a long run. Then again, you might say I've never stopped running—on the political track, at least.

Of course, part of coming to grips with any disability is recognizing that there will be some things that you just can't possibly do. But even people who are perfectly healthy can't do everything, or get everything done that they'd like to squeeze into one lifetime.

Perhaps the most important thing a person

with a disability can know is that he is accepted by others. For too long, disabled people felt like outcasts. But today, we are much more open about ourselves, and about our family members and friends with disabilities. You don't keep Aunt Minnie locked in her bedroom simply because she has lost a step or two, and may even be a bit senile. Today people are far more tolerant than they were when I was growing up.

Some people have hinted to me, "Well, it was easier for you to overcome the obstacles in your life. After all, you were a well-known senator."

True, but I wasn't well known that day on Hill 913. I was just one more second lieutenant from a small town in the American heartland. Like anyone else, I had to get up and take one small step after another to regain my life, and to do something meaningful with it. Sometimes I wonder what sort of person I would be today if I had not been on Hill 913 in April 1945. Thanks to my upbringing, I doubt that I'd have done things much differently. The drive to achieve was already there . . . it was in me long before I went off to war. Maybe I would not have successfully overcome my adversities in life had I not already had those qualities instilled within me, and had they not been tested in the aftermath of World War II.

One of life's great milestones is when a person can look back and be almost as thankful for the setbacks as for the victories. Gradually, it dawns on us that success and failure are not polar opposites. They are part of the same picture—the picture of a full life, where you have your ups and downs. After all, none of us can ever lose unless we find the courage to try. Losing means that at least you were in the race. It means that when the whistle sounded, life did not find you watching from the sidelines.

There certainly have been times over the years when I have grappled with the "why" questions. Why did this happen to me? Why just a few days before the war ended? The questions are unanswerable this side of heaven. It's taken me sixty years to come to grips with the toughest questions of life, and in some small way, this book is my answer.

But instead of wallowing in despair trying to figure out a dilemma for which I believe there is a much larger answer than I will ever know, I've chosen to focus on a different set of "why" questions.

Why waste time wondering what might have been? Why go through life feeling sorry for myself? Instead of asking "why me," why not do something to help others?

During all the years I served in the Senate, I kept in my desk the original cigar box in which the people of Russell collected donations to help pay for my operation. That cigar box and those donations reflected the culture that produced me, how we lived, and our common experience. Because I grew up in that kind of environment—a community that did instinctively what many people today regard as the government's role—I've always been suspicious of government intrusion into people's lives.

I wasn't merely against big government; I didn't want to lose what is so special about the American character—our big hearts, and the responsibility each of us has to his or her neighbors. We don't want to subcontract that out to politicians hundreds or thousands of miles away, not because they have evil intentions, but because we don't want to lose our American concern for one another. I know firsthand what compassionate conservatism is all about, because I experienced it firsthand from the people of Russell.

I know, too, that there are people who, for reasons far beyond their control, need help. But I also know that the greatest social program in the world is a neighbor who cares about you.

Where does that come from? It comes from the values that we learned in towns like Russell, the timeless values of faith in God, concern for one another, hard work, and optimism about America. It's not that I relish nostalgia or have idealistic images of an irrecoverable past. Quite the contrary; I believe in the future, I believe in possibilities—and the possibilities for every person in the United States are indeed great.

I once said that I was the most optimistic man in America. It was a phrase reminiscent of Franklin D. Roosevelt, who undoubtedly was the most optimistic man in America during his lifetime. Deprived of the use of his legs, he had been through his own personal hell yet continued to hope for the best. I could relate to that.

Today I am still an optimist. I believe that the greatest generation is today's generation. My optimism is based on the belief that anyone in America, whatever your race, age, or status, whatever your strengths, weaknesses, or disabilities, deserves an equal opportunity to succeed, and you can find that opportunity in America.

That's what we fought for in World War II, that's why I charged up Hill 913, that's what some of my friends bled and died for, and that's what I've lived for ever since.

Nearly sixty years after I headed up Hill 913, I concluded my speech at the dedication ceremony of the National World War II Memorial by saying, "It is only fitting that when this memorial was opened to the public . . . the very first visitors were schoolchildren.

"For them, our war is ancient history, and those who fought it are slightly ancient themselves; yet in the end, they are the ones for whom we built this shrine—and to whom we now hand the baton in the unending relay of human possibility. Certainly the heroes represented by the four thousand gold stars on the Freedom Wall need no monument to commemorate their sacrifice. They are known to God and to their fellow soldiers, who will mourn their passing until the day of our own.

"In their name we dedicate this place of meditation. And it is in their memory that I ask you to stand, if possible, and join me in a moment of silent tribute to remind us all that some time in our life, we have [been] or may be called upon, to make a sacrifice for our country to preserve liberty and freedom."

That's my story. It's only one soldier's story. Yet in a real way it is not merely a story about me; it is about Jim and Phil, Harry and Bob, and other guys who have suffered and struggled, but who still made a life. More important, they made a difference. They all had hills to climb, tough roads to walk. But they never walked alone.

My story could be told with variations by thousands of other men and women. My sacrifice is no different from that of millions of my generation, or of today's young Americans who will wear the scars as well as the medals of their service for as long as they draw breath. It's been said that God grants liberty only to those who love it and who are always ready to guard and defend it. My generation loved liberty as much as life itself.

Now it's your turn—your turn to run the race, your turn to climb the mountain, your turn to make a difference.

"Can't" never could do anything.

Epilogue

Today is January 20, 2005, and I should be at the inauguration of President George W. Bush. Instead, because of a cruel twist of fate, it's as if I have just been transported back in time to April 14, 1945, and I'm lying wounded on Hill 913. As it happens, I'm at Walter Reed Army Medical Center, in Washington, D.C.

On December 14, 2004, I underwent successful hip replacement surgery at the Hospital for Special Surgery in New York City. After an additional ten days of recuperation at Walter Reed, I traveled to Bal Harbour, Florida, one of my favorite places to rest and relax. I returned to Washington tanned and rested on Monday, January 10. About noon the following day, I was moving a suitcase when I lost my balance and took a nasty fall. I was on the bedroom floor, with blood streaming from my left arm and right eye, and a sharp pain engulfing my left arm—the better one.

Our housekeeper ran for assistance and found Mr. Walker, the apartment doorman, who lifted me off the floor. I called my office, and within an hour, Wilbert Jones, my driver and, more important, a trusted friend, was taking me to Walter Reed Medical Center. Between five o'clock and seven, I underwent surgery on my right eye. Around seven o'clock, I went home thinking everything was okay.

I didn't stay long, however. By ten P.M. I was feeling miserable. Near midnight, an army ambulance whisked me back to Walter Reed, where doctors discovered that I had multiple problems, including internal bleeding. The physicians quickly attempted to reverse the effect of a blood-thinning medication that had been administered to ward off clotting following the hip replacement.

Frozen plasma helped. But it took more than that. Great as their skills were, the doctors disclaimed responsibility for what happened next. They said that a Higher Power intervened to halt the bleeding and avert fatal complications.

The afternoon of Friday, January 14, I was

transferred from Ward 41 to Ward 72, otherwise known as the Eisenhower Executive Nursing Suite. Ward 72 is ordinarily reserved for presidents and first ladies, cabinet officers, members of Congress, foreign dignitaries, and others designated by the president or secretary of the army. It's a bit like a museum, with an old copy of the U.S. Constitution and an assortment of gifts from other nations on display.

My new surroundings conjure up some old memories. I'm back where I started, sixty years after Hill 913. I'm not paralyzed, but my left arm and hand are useless. With both arms out of commission, I must rely on others to feed me. I can't scratch my nose or go to the bathroom without assistance. It is humiliating. The past comes rushing back. At times I imagine I'm back in Percy Jones, or in the long halls of other army hospitals. The friendly doctors, nurses, and other hospital personnel offer words of encouragement and give perspective regarding occasional setbacks. It all seems so familiar.

The next few days telescope the struggle of my youth. Monday, January 17, is a turning point. I can move my left thumb and the fluid in my swollen arm and hand is subsiding. With some help from a strap placed under my body, I am raised to my feet by a couple of hospital attendants. Slowly, I attempt a few steps.

To combat infection in my left arm, the doctors have changed the medicine they've been administering. It works. My left knee is drained, and, presto—it works, too. Before you know it I can balance myself, more or less. My appetite returns. For the first time in a month a burger and custard taste good.

By Friday, the twenty-first, I celebrate another achievement—I can reach up, just barely, and scratch my face. Better yet, I can move my fingers and even bend my left arm a little. Over the weekend, I'm able to brush my teeth and use an electric razor.

Late one afternoon I get a call from my friend Tom Brokaw. "Life isn't fair," he says. I reflected for a moment, **Life, on the whole, has been more than fair to me**. Certainly I wouldn't trade my life for any other. I've been afforded opportunities, and accorded honors, that come to few people. My tests along the way have been only exaggerated forms of what millions of my generation went through in times of youthful hardship, on foreign battlefields, and in countless personal challenges.

To be sure, my war has been an unconventional one. Most of it took place long after peace treaties were signed, in hospital wards and operating rooms. If it left me scarred, it also prepared me for the longest walk of all. And, through it all, I've realized that I've never walked alone.